"I Touch the Future . . ."

"I Touch the Future..."

THE STORY OF CHRISTA McAULIFFE

Robert T. Hohler, *1951-*

Thorndike Press · Thorndike, Maine

B
M117 h
A-3

Library of Congress Cataloging-in-Publication Data

Hohler, Robert T., 1951-
  I touch the future —

  1. McAuliffe, Christa, 1948-1986.   2. Challenger
(Spacecraft) — Accidents.   3. Astronauts — United States —
Biography.   4. Teachers — New Hampshire — Biography.
5. Large type books.   I. Title.
[TL789.85.M33H64   1986]   629.45'0092'4   [B]   87-6532
ISBN 0-89621-811-2 (lg. print : alk. paper)

The excerpt from the song "Columbia," by Paul Candide
(Paul A. Fafard) is reprinted by permission of the author. The untitled poem
by Amy Heath is reprinted by permission of Marie Heath.

Grateful acknowledgment is made to the following for permission to reprint
previously published material:

W. Edward Dill: "Somewhere," lyrics to the senior class song by William E.
Dill, Jr., deceased March 10, 1980. Permission granted by W. Edward Dill
for the Estate.

Henry Holt and Company, Inc.: "Mourn Not the Dead," by Ralph Chaplin,
from *The Home Book of Modern Verse*, edited by Burton E. Stevenson.
Copyright 1925 by Holt, Rinehart and Winston. Copyright 1953 by Burton
E. Stevenson. Reprinted by permission of Henry Holt and Company, Inc.

I.H.T. Corporation: "High Flight," by John Gillespie Magee, Jr.,
New York Herald Tribune, February 8, 1942. Used by permission of
I.H.T. Corporation.

Warner Bros. Music: Excerpt from "You're Gonna Hear from Me," by
André Previn. Copyright © 1965 by Warner Bros. Inc. All rights reserved.
Used by permission.

Large Print edition available in North America by
arrangement with Random House, Inc., New York.

Cover design by Robert Aulicino.
Cover photography by Michael O'Brien/Archive.

For Lauren

PREFACE

I met Christa McAuliffe as a reporter for her hometown newspaper, the *Concord* (N.H.) *Monitor*. She was a social studies teacher from Concord High School, a finalist in a teacher-in-space contest conducted by the National Aeronautics and Space Administration, and I was sent to Houston in July 1985 to cover her. On the afternoon I arrived, we sat on a cool granite bench in the shade and talked about her home, her family, her fears, her dreams. We hit it off.

Christa won NASA's contest, and I shadowed her for two hundred days, traveling from Concord, Houston and Washington to her childhood home in Framingham, Massachusetts, and finally, in January 1986, to the John F. Kennedy Space Center in Florida. I woke her the night before her selection to ask if she had won the space odyssey. I curled up in the back of a station wagon the next day to ride from the White House with her in her moment of triumph. I sat with her an hour later when she called her husband, Steve, to share the news. I fidgeted, waiting for her to tire of me, but she never did.

Christa squeezed the most out of her final months. One day she was in Houston training to become the first private citizen in space, the next day in Washington exhorting the nation to recognize the importance of education. Keeping up with her demanded a quick pace, and occasionally the job was so consuming that my seven-year-old daughter, Lauren, complained she had lost me to a teachernaut.

"Everything's Christa this, Christa that," she said. "When's it going to be over?"

I was there when it was over, snapping pictures of Christa's parents as their proudest moment turned to tragedy — the deaths of their daughter and her six crew mates in the space shuttle *Challenger* on January 28, 1986. Even for those who knew Christa only a short time, a sense of loss lingered long after her death. But the pain was soothed by memories of Christa in life: memories of a teacher who asked to be nothing more than an ordinary woman on an extraordinary mission. This book is intended to preserve those memories.

It was written with deep fondness for

Christa and great respect for her family and the families of the entire *Challenger* crew. And written, too, with the hope that others will continue her mission to improve education for our children.

"I Touch the Future . . ."

CHAPTER ONE

In the quiet just after dawn, Christa McAuliffe kicked her feet out of her king-size bed and stepped into the thick, cool carpet. She groped through the dark to the picture window, and her stomach started to tingle as she tugged open the curtains on the day, maybe her glory day.

The morning sun burned soft and white. A swimming pool shimmered below her window, and the water seemed as inviting now as it had seemed the day before when she had dived in for a few moments of peace. But there was no time for that today, she thought. Christa and nine companions had an appointment at the White House.

They were teachers — social studies, science and language teachers, second-grade and third-grade teachers. When the last school bell rang in June 1985, they converged on Washington from as far as Texas and Idaho, each of them trying to convince the National

Aeronautics and Space Administration that they should pioneer the high frontier for the common man. Nearly 11,500 had entered NASA's teacher-in-space sweepstakes, and now there were 10. Soon there would be one, and as Christa laid out her clothes in a suite on the fourteenth floor of the L'Enfant Plaza, a posh hotel in the shadow of the capitol, she hoped to heaven it would be her.

She glanced out the window again as a breakfast cart rolled past her door. She thought about her colleagues down the hall, nine men and women who had helped her survive two weeks under the bright lights of NASA's test labs and the magnifying glass of the media. They had become friends under fire, and now, she remembered, they faced the terrible possibility of walking into the White House like contestants in a low-budget beauty pageant. The tingle in her stomach stopped.

Christa knew why they might be cast in a public spectacle. President Ronald Reagan had piqued the nation's curiosity eleven months earlier when he announced to students at a junior high school in downtown Washington that a teacher would be the first private citizen in space, a backseat rider on the space shuttle *Challenger* in January 1986.

As the news spread, millions of ordinary Americans began to believe that if a teacher could enter a world once limited to daredevil pilots, well, maybe, someday they could too. They were restless to meet the pioneer.

The network cameras were poised, but the president was in the hospital, a cancer patient, leaving Vice President George Bush to introduce the space-bound teacher. Bush's people intended to get the most for their media punch by keeping the winner's name a secret until the ten teachers stood before the spotlights and Bush informed the lucky one he or she would live forever in history — a television news clip for the ages.

Never mind the other nine, whose jaws would drop as they stood by the podium and pondered their futures as footnotes.

No, thank you, thought Christa, a high school social studies teacher from Concord, New Hampshire. Here we are, ten professionals who have invested a lot of energy and emotion, and now they want to make this an absolute circus. No, thank you. It's not fair to any of us.

She could see the videotape now. "And the winner is . . . Miss New Hampshire!"

"Miss who?"

She would unravel right there in the White

House. She knew it. Even worse, what if it was Miss Maryland? Or Mr. Massachusetts? How would she react then? How would the others react?

A few of them were already strained. At a birthday party for Kathleen Beres, a finalist from Baltimore, two nights before the White House announcement, Beres's boyfriend watched the teachers mill anxiously about the backyard barbecue. After dinner he tapped Beres on the shoulder.

"This group is really wound up," he said. "I mean, boy, there's only one normal person here."

"Oh?" Beres asked softly, expecting to hear her name.

"Yeah," he said. "Christa. The rest of you guys are wired for sound."

Only Christa found time to buy Beres a gift, a small plaque that said FRIENDS ARE UPLIFTING. On the back she wrote, "Dear Kathy, We already have a lifetime full of memories. Love, Christa." She also brought a birthday cake.

Like an angel from Madison Avenue, Christa seemed to be the perfect match for a space agency trying to renew America's romance with the heavens. At thirty-six she was just the right age — not too old, not too young. She was pretty but not too pretty;

16

brown-eyed with an engaging smile; and thick chestnut hair fell in curls to her shoulders. Her husband, Steve, a lawyer, was her high school sweetheart, and they had two lively children, Scott, eight, a lover of stuffed toy frogs and the *Star Wars* trilogy, and Caroline, five, a fan of Michael Jackson and spaghetti. They lived in a modest Victorian on a shady, all-American street.

Christa appreciated classical music, but she preferred Carly Simon, Bob Dylan and the Beatles. She taught Sunday school and spent the rest of the week dashing from Concord High School to theater rehearsals and co-ed volleyball games, one community activity to another, correcting school papers along the way. On summer nights she enjoyed a cold beer and a Ms. Pac Man video game. On a misty New England morning, she liked to rise early and jog with friends past the duck pond near her house. Full of life, she had no time for people who believed that you die a little every day.

She was even a Girl Scout. Who better to sell the wonders of space than a woman who once sold more Girl Scout cookies than anyone in her neighborhood? And she still had the touch. She was bold, charming and convincing, and when she said in her teacher's

17

voice "I want to prove that space is for everyone," people believed her.

But did NASA believe her? Did NASA want her?

The night before the White House announcement, she sneaked away from the hurly-burly of a hotel banquet with Peggy Lathlaen, a finalist from Friendswood, Texas. They walked several blocks to a small café near the Smithsonian Institution and settled at a corner table where they ordered drinks and talked about their chances.

Overnight, it seemed, the odds of winning a six-day ride on *Challenger* dropped from nearly 11,500 to 1 to 113 to 1, and now to 10 to 1. Everyone from the bellhops at their hotel to the reporters covering the story had bet on a winner, but Christa was stumped. Each of the candidates was so articulate, so creative, she said, so perfect. Why not just pick a name out of a hat?

She told Lathlaen she had done her best to convince NASA that she could "humanize the technology of the space age" by showing the world that "there are real people up there." She had passed the high-tech medical tests, survived a series of simulated space sensations and satisfied a psychiatrist that she was willing to risk her life for an adventure she described as the ultimate field trip. And if she was lucky enough to win NASA's talent search, she knew her family was ready to live without her for a year of training and public relations work. She worried only about her husband's diet: when she was away, he had a peculiar habit of trying to survive on cornflakes.

Still, Christa slept easily that night. Ignoring the advice of a NASA official who told each of the finalists to prepare a victory statement, instead she called her husband, her parents, her friends and relatives across the country. Then she settled into her king-size bed, content that she had made the most of this crazy summer vacation. Tomorrow night, win or lose, she would sleep at home, and that was as comforting as a summer breeze.

Across the river in Maryland, NASA's chief administrator, James Beggs, lay in bed and wondered about the recommendation a selection committee had handed him the night before. He'd told them he would sleep on it, and by breakfast the next morning he had decided to accept that recommendation.

Searching for a lucky charm, Christa snapped on her mother's jade bracelet and a silver watch her late grandmother had given her. A string of pearls, a gift from Steve, hung from her neck, glistening on a summer sweater her sister Lisa had given her and which Christa wore with a beige pleated skirt, another gift from Lisa. She pulled on a bright yellow jacket her mother had bought, slung a tiny black purse over her shoulder and joined the other finalists to plot an uprising.

Their target: the White House ceremony they feared would turn into a soap opera. Christa was not the only angry teacher: all ten believed they had invested too much emotion to have it splash back in their faces right there on national television.

Their strategy: if NASA refused to defy the vice president's people and reveal the winner beforehand, they would boycott the ceremony. No dignity, maybe no teacher in space.

The message rose quickly through NASA's chain of command, landing in Beggs's lap by midmorning. A year earlier, Beggs had trumpeted the selection of a teacher as the first private citizen in space by saying, "This agency lives and dies by whether we can attract top talent and keep kids interested in the space program."

He had no intention of clashing with teachers now. Yes, he said, they would learn who the winner was, but only on the condition that absolutely no one leaked that name to the press. If everyone kept silent, maybe the White House would not discover the little shift in protocol. The last thing Beggs needed was to hear the winner's name on the radio as he drove the ten minutes to the White House.

So the finalists met Secretary of Education William Bennett, who offered to substitute one day at the school of the winning teacher. They were then tucked away in a conference room on the seventh floor at NASA's white marble headquarters. Lunch was delivered, the door closed and everyone in the room believed no one else in the city knew where they were hiding.

Minutes later a reporter walked in. "Any word yet on the winner?" he asked.

Suddenly the ten teachers and their two NASA protectors were on the run, whispering and tiptoeing through the tiled corridors, their movable feast on a squeaky food cart, searching for another hideout.

"It was crazy," said Beres, who was awakened at 4:00 A.M. by a *USA Today* reporter looking for the scoop. "It was like a scene in a Peter Sellers movie."

As they dashed from room to room, Lathlaen, a tall, gentle Texan who was a bit unnerved by all the commotion, stayed at Christa's shoulder. Christa had kept Lathlaen from panicking during a frightening moment a week earlier when Lathlaen's oxygen mask had begun to leak during an altitude test at the Johnson Space Center.

"Something about Christa was very comforting," Lathlaen said. "Her eyes said peace and calm."

The name of the first private citizen in space was hot news, particularly in cities that claimed a teacher in the final ten. The *Washington Post*, with finalists from Virginia and Maryland, sent its top NASA writer and a White House reporter after the name. In Christa's hometown, the *Concord Monitor* tried everything: the reporter they had sent to Washington, friends at the *Post*, friends in the Senate, Christa's husband. They tried them twice, sometimes three times, with no success.

The answer was now in room 7002 of NASA headquarters, an unoccupied office overlooking a Holiday Inn. Amid a clutter of boxes and dusty computers, the teachers sat on couches, finishing roast beef and turkey sandwiches and washing them down with lukewarm sodas. They had made their final phone calls and now chatted, waiting to learn the winner's name before being whisked to the White House.

At Christa's childhood home in Framingham, Massachusetts, her father, Ed Corrigan, waited in the living room to hear whether or not she had won. As he sat there he remembered the nights many years earlier when he had played show tunes on the family piano and Christa had sung for him. He started to hum her favorite song, André Previn's "You're Gonna Hear from Me."

He closed his eyes and imagined Christa's singing under the light of the chandelier:

Move over sun and give me some sky.
I've got some wings I'm eager to try.
I may be unknown, but wait 'til I've flown.
You're gonna hear from me.

Remembering the song at that moment "must be a good omen," he told his wife, Grace.

When Christa had told her sister Lisa Bristol about applying for the two-million-mile ride in space, Lisa had said, "That's nice. What else is new?" But now Lisa was calling her parents in a panic to tell them that pink and black were the lucky colors of the day. Wear pink and black, she said, and Christa would fly.

In room 7002, Christa sat next to Niki Wenger, a finalist who taught gifted junior high school students in Parkersburg, West Virginia. Nearby sat Ann Bradley, a NASA executive and chairman of the selection committee. Her job was to tell the teachers who had won, and she was waiting for the right opportunity. It was a little after noon on July 19, 1985.

As Bradley sipped from a can of Coke, Christa told Wenger about her husband's strange interlude as a single parent. Steve believed his unusual housekeeping methods made him the perfect star for a sequel to *Mr. Mom*, and he joked about writing the *Space Husband's Cookbook*, featuring a chapter on "Snickers: The Forgotten Breakfast Food." When Christa started to tell Wenger about Steve's romance with cornflakes, Bradley lowered her Coke can and cut her off.

"Excuse me, Christa," she blurted out, "but you better tell your husband to stock up on a lot more cornflakes."

Stunned, McAuliffe felt her shoulders sag. Then her chin fell in a suspended moment of disbelief.

"Excuse me?" she said to Bradley.

The other teachers looked at each other like puzzled students.

24

"What . . . is . . . in . . . Ann's . . . Coke?" Beres asked no one in particular.

Bradley kept staring at Christa. "You're the one," she told her.

"I'm the . . . ?"

"You're going," Bradley said. "You're the one going up in space."

Christa threw her hands to her cheeks and waited for her heart to accept what her brain was trying to tell her — in fewer than two hundred days, if this wasn't a dream, she was going up in space.

Christa was not the brightest of the ten finalists. One of them was a prize-winning playwright and poet, and another had been invited by the French government to study language, literature and culture there for a year. Most of them had graduated from schools more prestigious than Christa's alma mater of Framingham State College. One had even graduated Phi Beta Kappa from Stanford University.

On paper, some of them seemed to push Christa to the back of the class. There was a former fighter pilot, a film producer and a woman who, among other adventures, had climbed the Andes and Himalayas and crossed the Atlantic in a thirty-one-foot sailboat. Several of them knew much more about space and science than Christa knew, and the projects most of them had proposed for the six-day journey made Christa's idea of keeping a diary look rather ordinary.

Which, of course, was the difference. Christa was the girl next door, and more. No other finalist matched her potential for getting NASA's message across to so many people.

Christa sat in shock as the others waited in silence, some in relief, most in disappointment. No one moved until Lathlaen recognized the wisdom of NASA's choice, and said, "Christa, sure, of course — Christa." Then she rushed across the carpet and gathered Christa in her arms as the eight others formed a crescent behind them.

Her cheeks wet with tears, Christa embraced each of the other teachers before she broke away to ask Bradley if she could call Steve, a request Bradley and other NASA officials denied several times during the day.

"Boy, I'd really like to talk to the man," Christa said softly again and again.

No stranger to success, Christa had been crowned summer princess of the Saxonville playground in Framingham at the age of six. She earned a drawerful of Girl Scout badges, made the sports teams and drama groups she tried out for, decided in high school on the man she wanted to marry and embarked on an adult life in which she knew what she wanted and went after it. But this — something about this was a bit harrowing.

She needed help, and with Steve out of reach she needed her nine friends to rally around her on the day she faced the nation. As she started to explain that she might need them for a much longer time, Bradley picked up on another cue.

"Excuse me," she said, "but Christa's really going to need a support person over the next few months, someone who can serve as sort of a cheerleader. So Barbara Morgan, you're the alternate teacher in space."

A second-grade teacher from McCall, Idaho, Morgan graduated from Stanford, taught science in Ecuador and remedial reading on the Flathead Indian Reservation in Montana. Later, she settled in the Salmon River Mountains of Idaho with her husband, Clay, a fiction writer and a firefighter for the U.S. Forest Service. At the moment, Clay was recovering from an accident in which a burning tree fell on his back, bruised his ribs and knocked him onto a branch that pierced his jaw and broke two of his teeth.

There was no glitz in Barbara Morgan's world. She rarely wore makeup, her clothes were simple and tasteful, and for fun on a Saturday night she played the flute in a small chamber orchestra. The finalists called her a cheerleader because of her uncanny knack of knowing exactly when and how to lift the spirits of a slumping colleague.

So Christa and Barbara, the girl next door and the cheerleader, would represent the five billion people of the world who had never entered space. Awed by the assignment, they found their female colleagues and retreated to the ladies' room.

"By the way, did you write an acceptance speech?" Beres asked Christa as the teachers gathered like teenagers at a pajama party.

"Oh m'gosh, no," Christa said. "I forgot."

She had also forgotten her hair brush and lipstick, which she borrowed from her friends before the ten of them and two NASA administrators squeezed into a pair of gray government station wagons for a ride to the White House. Christa composed her victory statement on the way.

At once Christa seemed to be a traditional mother and a daring pioneer. The space shuttle never tempted her as a billion-dollar carnival ride. Christa loved her husband

and children too much to dash away for a year of high-tech fantasies. She went to Washington because she wanted to teach about a new frontier, about personal challenge and about the power of her much-maligned profession.

"Imagine me teaching from space, all over the world," she said in her heavy New England accent. "Touching so many people's lives — that's a teacher's dream."

Her first big chance was only blocks away, but as she rode down Pennsylvania Avenue past lunch-bound office workers, teenagers weaving through traffic on bicycles and shirtless construction crews eating sandwiches in the shade none of that crossed her mind. She thought instead about her nine friends. In their two-week gauntlet of mental and physical exams, they had teased and tested each other, encouraged and comforted each other. Now she felt as if it were time to say good-bye on the last day of Girl Scout camp. Christa wondered if she could stand before the cameras and keep from crying.

"Don't lose your cool," she said to herself. "Don't let your students see Mrs. McAuliffe in tears."

At the last minute the White House ceremony was switched from the Oval Office to the Roosevelt Room to accommodate a swelling legion of reporters and photographers. Cable News Network planned to carry it live; CBS, NBC and ABC were saving spots near the top of the nightly news. The *Washington Post* and the *New York Times* wanted it on the front page, as did smaller papers from Portland, Maine, to Portland, Oregon. The White House press room was standing room only.

Just after 1:00 P.M., Robert McFarlane, the president's national security adviser, was briefing reporters on the SALT talks when a door opened at the front of the press room and a member of the vice president's staff signaled that the teacher-in-space ceremony was about to begin. Within seconds, most of McFarlane's audience disappeared.

They swarmed, many of them with cameras on their shoulders, to the Roosevelt Room, a small conference area where the president held his daily staff meetings and sometimes swore in cabinet members. Before Richard Nixon dedicated the room to Theodore Roosevelt, it was known as the Fish Room, the place where presidents from Franklin Roosevelt to John Kennedy hung their fishing and boating trophies. In Franklin Roosevelt's time, visitors sat in the Fish Room to cool off before meeting the president. His staff called it the Morgue.

Photographers squeezed around the Chippendale furniture as the teachers filed into the room with Bush, Beggs and Bennett, flashes firing in their faces.

"Who's the winner?" reporters asked.

The teachers smiled, but, to the relief of Beggs, none of them answered.

For two weeks, NASA media experts had lined up the teachers in alphabetical order or made sure they divided the men and women. This time the teachers were on their own, and Christa stood closest to Bush, a red rose given her by Wenger sprouting from her yellow lapel.

As Beggs began to speak, Christa knew in an instant that she would be more famous than Eugene Cernan, the last man to leave footprints on the moon. She glanced over her right shoulder at Lathlaen, her eyes wide with alarm.

"I'm right here," Lathlaen said, and Christa grabbed her hand.

Beggs said it was not easy picking a winner, "but I'm confident that when the shuttle lifts off, our winning candidate will soar with it right into the hearts and minds of young people around the country, indeed, around the world."

Around the world? Christa thought, squeezing Lathlaen's hand harder.

She laughed when Beggs turned to the teachers and said, "Just as we have gotten to know you inside out . . . "

Inside out, indeed. Would they ever forget their visits to NASA's proctologist, a visit that prompted Christa to admit she was learning about parts of her body she never knew existed?

Then her laughter trailed off as Bush unfolded his notes and prepared for a moment Christa had dreaded at dawn. Judith Garcia, a finalist from Alexandria, Virginia, stroked her back.

Surprising no one, Bush needed fewer than four sentences to talk about the "teacher with the right stuff," and Christa couldn't help but giggle as she squinted into the television lights. The vice president continued quickly, introducing Morgan, shaking her hand and presenting her a trophy, a bronze statuette of a teacher pointing a student toward the stars. Then he turned a page and raised his voice a notch.

"And the winner, the teacher who will be going into space, is Christa McAuliffe, who is . . . " He glanced at the group and guessed instantly that the woman in the bright yellow jacket, the one clutching her colleague's hand

and flashing a smile wider than Chesapeake Bay, must be the winner.

"Is that you?" he asked, realizing at once that she would survive the historic announcement without a hug from her vice president.

Christa shook his hand, cradled a trophy like Morgan's and waited for Bush to step back from the podium. This was it, the moment she needed her dignity. Christa leaned toward the microphone. Her chin trembled.

"It's not often a teacher is at a loss for words," she said, her voice quivering. "I know my students wouldn't think so."

Then, quickly but tenderly, she said, "I've made nine wonderful friends over the last two weeks, and when that shuttle goes, there might be one body" — sobbing once, she raised her finger to her lips to steel herself — "but there's gonna be ten souls I'm takin' with me."

She did it, right there in the Roosevelt Room. NASA's angel had wings.

"How does it feel?" asked a reporter, one of dozens awaiting her outside the West Wing of the White House.

"I'm still kind of floating," she said. "I don't know when I'll come down to Earth."

Surrounded by friends and co-workers in

his law office on Main Street in Concord, Steve McAuliffe had held his Dictaphone next to a radio to tape the announcement. When he heard Christa had won, his chin fell as hers had fallen. The pop of champagne corks brought him around, and he shook his head, repeating, "Unbelievable . . . unbelievable . . . unbelievable." Then he laughed triumphantly and called home, where Scott — Steve called him Bunky — had huddled by a radio with Steve's mother.

"Hey, Bunky boy, what d'ya think?" Steve asked.

"Does this mean we get to Florida?"

"It does indeed mean we're going to Florida."

Caroline had passed up the announcement to swim with friends at the Concord Country Club pool. When reporters reached her later, she told them what she had heard her brother tell other reporters earlier: "I feel like going up, too."

When they asked her why, she giggled. "Because I think it's fun."

A few blocks down Main Street from Steve's law office, a disc jockey announced the news over a public address system, and scores of shoppers at a sidewalk festival threw up their arms and cheered. In Diversi's Market, the owner, Mary Ann Lakevicius, broke down in tears.

"I'm so happy for her," she said.

The Associated Press sent an urgent bulletin across its national wire two minutes after the White House ceremony: "Vice President George Bush today named Sharon Christa McAuliffe, a teacher at Concord High School in Concord, N.H., as America's first citizen passenger scheduled to go into orbit aboard the space shuttle."

The top reporter at the *Concord Monitor*, David Olinger, had spent ten years writing about triumph and tragedy. Now he sat at his desk in the newsroom and cried.

"It was an amateurish thing to do, but I was so proud," he said. "*Our* teacher had won."

Meanwhile, Christa stood on the North Lawn of the White House, clutching her Oscar-size trophy and melting an icy press corps, firing off perfect fifteen-second sparklers for the television and radio people and feeding the print reporters snappy anecdotes and snippets of philosophy that left their editors smiling.

On cue, she talked about looking down from the shuttle on "Spaceship Earth" — a Disney concept — imagining a planet where no differences divided blacks and whites,

Arabs and Jews, Russians and Americans.

"It's going to be wonderful to see us as one people, a world with no boundaries," she said. "I can't wait to bring back that humanistic spirit."

"But won't the flight frighten you?" a reporter wanted to know.

"Oh no," Christa said. "It's not like the early missions when the astronauts had no control of their destiny. Now they can make emergency landings or orbit the Earth once before landing. There's a lot less to worry about."

When harried television producers pressed her, she handled them with humor.

"Hold up the statue," one of them said. "Turn this way. . . . Okay, now step over here. . . . Great, just snap on this mike. . . . Now count down from ten."

"Count down from ten?" she asked. "Then what do I do? Blast off?"

Seconds later, NASA was under attack and Christa was fighting back.

"What about the fact that there were no blacks in the final ten?" asked a feisty blond woman waving a microphone. "What does that tell millions of black schoolchildren across the nation?"

"Oh, goodness," Christa said, looking at

the reporter like a student whose homework was overdue. "There was no place on the application for race, sex or age. It was totally color-blind."

Christa was about to continue but David Marquart, a finalist from Boise, Idaho, swept her into a final congratulatory hug.

"Good luck, kid," he said.

"Oh, Dave, I'm still pinching myself," she said. "My feet might not touch the ground for days."

A new set of reporters arrived, eager to know about the space teacher and her family.

Yikes! Her family!

"I can't wait to get home and see everyone," she said. "My children are anxious to see me, my parents are coming up and I promised them I'd take the plane that gets in at nine-thirty tonight . . ."

Just then, a NASA man wearing a Ronald Reagan tiepin pulled her aside and whispered in her ear.

"Well, o-o-o-kay," she said, her voice dropping a decibel. "It looks like I'm not gonna make that plane."

At the moment Bush announced that she would begin the common man's migration into space, Christa's life was no longer her own. She belonged to NASA, and nothing about the high price of celebrity hurt her more than losing her family, even for a while. She figured Scott, her "space-age child," could handle it. She had already promised him she would take up his favorite stuffed frog, Fleegle, and wave to him when she orbited over Concord. But she knew Caroline was too young to understand.

Even Scott tried sitting Caroline down and setting her straight.

"Look, Caroline," he said, "this is a once-in-a-lifetime opportunity."

But Caroline, who usually valued her brother's opinions, said it still wasn't fair.

By now Christa's escorts were ready to leave the White House — quickly. The teachers informed the press after the ceremony that, yes, they had been told who the winner was before they arrived in the Roosevelt Room, and one of Bush's aides, a tiny woman with a booming voice, suddenly came undone.

"I hope you know you've embarrassed the vice president of the United States," she told one NASA official, then another.

Before matters worsened, the teachers and the NASA people hopped into their station wagons and retreated to headquarters.

"You were terrific," Lathlaen told Christa,

leaning over to offer her a fresh rose as the car weaved through traffic. "Weren't you nervous at all?"

"I was just so afraid I'd start crying," she said. "Other than that, it was like a day at school."

In the lobby at NASA headquarters, Christa was swarmed by executives and cleaning women, clerks and engineers, knots of people seeking an autograph or a handshake. Some of them rode in an elevator with her and followed her to the door of a television studio, where NASA's public relations people waited to powder her face for another round of interviews.

"More powder on her nose, cheeks and forehead," a producer told an assistant. "She's shining."

"Oh, gosh, I've never had so much makeup on in my life," Christa groaned. But she needed more.

"Try using a brush," the producer told the assistant. "We've got to get that shine off her."

"Forget it," Morgan said. "You'll never get it off. She's too excited."

At 2:50 P.M., more than two hours after walking into the White House, Christa made one more simple request of NASA officials. "I still haven't talked to my husband yet," she said, losing some of her shine. "At some point, I'd really like to talk to the man."

Not yet, a public affairs person said, and while Christa sat waiting before a large mural of *Challenger* touching down from its inaugural flight, a reporter asked her how she expected to keep her instant celebrity in perspective.

"Well, I see this as a very, very exciting time in my life," she said, "but I have thirty-six exciting years behind me and thirty-six more ahead of me. It's not like I'm never going to get back to Earth again."

A half hour later, she closed the door to a small office, settled behind a steel desk and started dialing her husband. A portrait of nine women astronauts hung on one wall; on another was a poem: "Always Have a Dream."

She called Steve's law office, but the line was busy. She called home, but the line was busy. She had no luck at a neighbor's house, either. Finally, the phone rang at the law office.

"He's in a meeting?" Christa said to the secretary. "Well, is there any chance I could talk to him for just a minute?"

Her strongest supporter, Steve had seemed

so enthusiastic the night before about her chances of winning that Christa had feared she might disappoint him.

"What if I *don't* get it?" she had asked, worried he might unravel in defeat. "You might not be able to practice law for the next ten years."

So now, as he came to the phone in her moment of triumph, Christa expected a word of congratulations, a few kind words, anything, really, would have been nice.

"What happened?" Steve asked.

"What do you mean, 'what happened'?" Her eyes widened. "You mean you didn't hear? You don't know? C'mon, you know. You *must* know."

Steve played dumb, and Christa, a bit puzzled, played along. "Are you kidding?" she said. "You mean you have no idea? You must know. Ste-e-e-eve, c'm-o-o-n."

Finally, he relented, and she threw her head back in mock disgust. But before Christa could protest his little charade, he asked her when she would be home.

"I'm not sure," she replied. "All I know is they promised me I'd get home sometime tonight."

"I hope so," Steve said, and he told her about the party he had organized for later that evening. "By the way, do you have any plans for tomorrow?"

"Plans for tomorrow?"

She could see the dishes and the dirty socks and . . .

"I hope not," he said, "because I committed you to the Lions Club parade in the morning."

"Parade? In the morning? Steve, I'm still pinching myself."

Christa told him about the moment she learned she was selected, and she laughed when she explained that his cornflakes diet was now a matter of public record. And once Christa laughed, she couldn't stop. All the suspense, all the strain of the day, all the joy converged in one long, hearty laugh.

"I still can't believe this," she said, catching her breath.

Christa leaned back in her chair and told Steve about her wardrobe — jacket, sweater, bracelet, watch and pearls. Her jacket and skirt were wrinkled from all the hugs, she said, but she was grateful that her friends had shown such strength and grace in her moment of triumph.

"But wait a minute," she said suddenly. "Who's gonna teach my course?"

She had developed a course on the American woman, a course that no one else had taught. Steve told her not to worry, that everything would work out well. Then he told her how proud he was of her. Her eyes watered, and she paused a minute. Then Christa glanced at the portrait of the women astronauts.

"Steve," she said, "this just isn't possible, is it?"

CHAPTER TWO

They lived in a one-room apartment in Boston not far from Fenway Park and a long way from the Ritz. Money was tight, much too tight, so the baby, Sharon Christa Corrigan, slept in a car bed while the parents slept on a couch. Soon their money was so tight they had no apartment.

The parents, Ed and Grace Corrigan, had met at Crosby High School in Waterbury, Connecticut, in 1940, the year Pan American Airways offered the world's first regular transatlantic flight, a quick hop from New York to Lisbon in twenty-six and a half hours. They had been friends, never more than that, and they had lost touch when Ed joined the Navy after graduation.

A world war later, Ed and Grace met again. It was 1947, New Year's Eve in New Haven, and they spotted each other on the street, Ed with his neon-blue eyes, Grace, a slender brunette with an electric smile and quick step.

They made a date to meet again, and they were married seven months later.

Children would have to wait, they agreed. No question about it. Ed was entering the school of industrial management at Boston College, Grace had no job, and the last thing they needed was to keep a baby in diapers while they tried to keep themselves out of debt. But their plans went awry and their first child, brown-eyed and bound for glory, arrived at the Boston Hospital for Women on September 2, 1948, a week before Ed started his sophomore year.

That was the first problem. The second was her name. Convinced all first children were boys, the Corrigans intended to honor Ed's Scottish heritage by naming theirs Christopher.

"Having a girl never entered our minds," he said. "Imagine our surprise."

On the night Grace missed the *Hallmark Playhouse* on the radio to deliver their eight-pound, thirteen-ounce girl, Ed rushed out and bought a book of names. They started with Abigail, ended with Zorah and settled on Sharon Christa because, well . . .

"Sharon sounded good," Ed said, "and Christa was the Scottish feminine form of Christopher. Simple as that."

But Sharon never stuck, and for the first nine months of her life they called the baby Sheri Christa. Then when her hair turned from brown to blond and she began to look Scandinavian, they dropped Sheri.

Why?

"Christa sounded Swedish," her father explained. "So we said, 'Why not just go with it?' "

They went with it, and Christa believed for twelve years that her parents had given her only a first name. When people asked her her middle name, Christa made one up.

"Mary, Margaret, Jane . . . I had one for every occasion," she said.

Every occasion but one — her eighth-grade church confirmation. Preparing for the ceremony, Christa walked to the rectory one day to pick up her baptismal certificate. On the way home, she opened it. She checked the birth date — no problem. Nothing unusual about the date of her baptism, either. But Sharon? She didn't know any Sharon, and she pivoted to run back to the rectory to tell the nun she had the wrong certificate. Then Christa read on.

"We've been meaning to tell you about that," her mother said when Christa burst through the door, flushed, waving the certificate in front of her.

"Imagine my surprise," she said.

She signed her name "S. Christa Corrigan" after that, prompting several people to believe the S was an abbreviation for a religious title. When letters arrived for "Sister Christa Corrigan," she laughed, but she never dropped the S.

The third problem for the Corrigans was diapers. They could afford so few of them that Grace washed them by hand and ironed them dry on the kitchen counter, an inconvenience that soon was dwarfed by a more menacing matter — illness.

Less than a mile from their neighborhood, a chain of yellow-brick buildings on Boston's Park Drive populated by Brahmins and Bohemians, mailmen and musicians, young couples and college students, was Children's Hospital. Christa visited often, suffering first from asthmatic bronchitis, a condition that caused several sprints to the emergency room and many fitful nights for her parents.

"We worried about going to sleep," said Grace, who had never held a baby before Christa was born. "We were scared to death she wouldn't be breathing when we woke up."

The asthma cured itself, but soon another ailment — infant diarrhea — took Christa from a car bed to a hospital bed, where she lay for twenty-eight days under the glare of a stark, white light, smiling occasionally at her parents as intravenous tubes fed liquid nutrients into her head. Christa was six months old, and she weakened daily.

"We thought we were going to lose her," Grace said.

A new antibiotic saved the baby, but the medical bills cost the Corrigans their home. Even their Spartan life became impossible, so Grace returned with Christa to Connecticut and Ed moved in with college friends, expecting to live apart from his family until he graduated. Money was so tight they didn't need a bankbook to know "we had peanuts."

Then their luck turned. Ed had a friend who knew Boston Mayor James Michael Curley, a politician who pulled more strings than a puppet troupe. When Ed learned the city was creating low-income public housing out of barracks that once sheltered Italian prisoners of war and later black American soldiers, he asked his friend to help him unravel the bureaucratic tangle. Ed asked him to approach the mayor.

A legendary benefactor of Boston's poor, Curley rarely disappointed his friends, and for twenty-three dollars a month the Corrigans had a home, a cold-water flat with no tub and thin walls on Columbia Point, an isolated pit of land that juts into Dorchester Bay. There was a bedroom for Christa and a living-room couch for her parents. A cold rain whipped off the harbor on the day they moved in.

"We were delighted," Grace said. "We were a family again."

Christa learned to walk when she was ten months old, her brown eyes gleaming and her long blond curls bobbing as she darted about the tiny apartment. She never spoke baby talk, always complete sentences, and she held her first conversation when she was a year old. She recited nursery rhymes before she was one and a half.

"We thought all kids did that," her father said. "It took awhile before we realized how special she was."

A neighbor noticed, though. When Christa was two and a half, a woman in the next apartment told the Corrigans they ought to check on their daughter more often at night. The woman said she had often heard Christa's tiny voice through her bedroom wall long after she should be asleep. Puzzled, Grace and Ed stayed up later than usual that night, and when they opened Christa's door they found her kneeling on a little red rocking chair. A record was spinning and she was singing along, rocking gently in the glow of her night light.

"She never had much time for sleep," Ed said. "When she wanted something, she went right after it."

Even new frontiers tempted Christa at an early age. She seized every opportunity to stray toward the rocks on the edge of the harbor or the tall buildings of the Boston skyline. Most of the time her dog, Teddy, a tiny mongrel, retrieved her before she stepped off the cement sidewalks around the housing project. But one Sunday morning when she was three, Christa rode away on her tricycle before Teddy or her parents knew she had gone. Pedaling several blocks to Columbia Road, a main thoroughfare into the city, she wheeled into the street and headed downtown, cars approaching from both directions.

A neighbor spotted her, but Teddy, bounding along the pavement, reached Christa first. Yelping and running circles around the tricycle, the dog stopped traffic, latched on to Christa's pants and tried to tug her back to the sidewalk. With the neighbor's help, he succeeded.

"We never did figure out where she was going," Ed said.

Christa was so active, so adventuresome for her age that her parents sometimes had trouble finding a babysitter for her. Even her aunt declined.

"She scares me," the aunt said. "She may be two and a half, but she makes me feel like a child."

All the maturity in the world couldn't cure Christa's motion sickness, however. Car rides made her queasy, and carnival rides made her downright ill.

"She even turned a little green on the hobby horse," her mother recalled.

The lean years ended with Ed's graduation. From Boston, the Corrigans returned to Waterbury with Christa and the second of their five children, a six-month-old boy named Christopher. Ed started an accounting career and Christa entered modeling school, where she was selected to appear on a local television fashion show at the age of four.

But her life in the footlights was fleeting. Ed took a job a year later as an assistant controller at Jordan Marsh, a Boston department store, and the family moved into a modest white ranch house on a winding street in a quiet middle-class neighborhood of Framingham, far from the grimy pavement near Fenway Park and Columbia Point. A bronze eagle hung on the front door, and in the main hallway hung a picture of Ed holding Christa, pig-tailed and wearing a white party dress, at his graduation.

Christa gave up modeling in Framingham and settled for wearing the crown as summer princess at her neighborhood playground.

"She was always a winner," said Anne Malavich, her best friend since childhood. "That's not unusual. The unusual thing is that I can't think of anyone who didn't like her."

Even when children called her "chipmunk," teasing her about her two protruding front teeth (she corrected them with braces when she was twenty-six), Christa kept her composure.

"She had an amazing amount of patience," Malavich said. "Maybe it came from being the oldest child. I'm not sure, but I never saw her lose her temper."

Christa fashioned herself after her mother, a woman of compassion and community spirit. Grace taught nursery school and served in several church groups. She was active in local politics, started a PTA chapter at the neighborhood school and organized a Brownie troop that launched Christa's career in the Girl Scouts. In her spare time, Grace painted in a makeshift studio at home.

"If I slow down, I'm no good," she said in January 1986. "I hate getting older because there's no time to do all the things I want to do. Life's too short."

With the same passion, Christa set aside time after Girl Scouts for dance, voice and piano lessons, Christian doctrine classes and sports practices. She looked after her brothers and sisters — Christopher, who was a year younger than she; Steven, who was three years younger; Betsy, who was nine years younger; and Lisa, who was born ten years after Christa. She sewed, cared for Teddy and still found a half hour on Friday nights to watch her favorite television show, *Superman.*

Her parents never pushed her, but she pushed herself, sometimes too hard. When her second-grade teacher saw Christa strain to the edge of frustration in penmanship, she spoke to Christa's parents.

"Christa has to accept the fact that she can't do everything perfectly," her teacher said. "I'm afraid she'll be disappointed if she spends the rest of her life as a perfectionist."

Later that year — October 4, 1957 — the

Soviet Union won the first leg of the space race by firing *Sputnik*, an unmanned satellite the size of a beachball, into orbit at twenty-five times the speed of sound. Seven months later, *Sputnik II* hurtled into space with a dog aboard. Then, on April 12, 1961, a Soviet Air Force lieutenant, Yuri Gagarin, became the first human to enter space.

The only American to puncture the Earth's atmosphere had been a chimpanzee.

"It's our worst humiliation since Custer's last stand," said a college professor in Pittsburgh.

Christa's idol, President John F. Kennedy, felt the national pangs of inferiority and pressed NASA to send a man into space. Less than a month later, a little after 9:30 A.M. on a Friday in May, Christa sat with her classmates in a cafeteria at Framingham's Lincoln Junior High School, counting down the final seconds to America's first manned space flight. As a giant Redstone rocket jumped about the screen of a black-and-white television showing Alan Shepard's brief trip into the heavens, Christa scribbled her impressions in a notebook. Later she told a classmate that she, too, would like to ride in space.

Astronautics was not a career choice, however, for women in the early sixties. They were encouraged to become teachers or nurses, secretaries or stewardesses, nuns if they happened to be Catholic. "Woman astronaut" was a contradictory term, so when Kennedy announced in Houston three weeks after Shepard's flight that America "should commit itself to achieving the goal, before the decade is out, of landing a man on the moon and returning him safely to Earth," Christa knew exactly who would make the journey — a man. Still she reveled in the promise of the high frontier.

She read *Profiles in Courage*, Kennedy's book about testing the limits of your potential, and, instead of kites, she played with toy satellites that landed on her neighbors' roofs. She watched on television as the astronauts rode away from Cape Canaveral on a trail of fire. She saw them plucked out of the Pacific upon their return and showered with ticker tape by adoring crowds on Broadway. She saved feature stories from *Life* magazine and reeled off without a hitch the names of the original seven astronauts.

But space was not her consuming passion. Like most children, she cared more about pleasing her parents, particularly a father who doted on her. As the oldest child, she had a special place in his heart. She cherished his attention, and while her friends watched *Father Knows Best* after dinner, Christa sang for her father as he played the baby grand piano in their living room. She never protested when he invited the neighbors to listen, or when he encouraged her to enter the town's talent show each year. Her greatest regret in life, she said later, was giving up piano lessons. She always wanted to play as well as her father.

Christa loved it when he took her to Boston on her birthday to see *South Pacific* on the big screen, or when he drove the family into the city each New Year's for dinner at Jimmy's, a fashionable seafood restaurant on the waterfront. The family didn't travel much — Christa grew up "like probably half the kids in the country . . . thinking states were different colors" — so she relished the summer trips he organized to Cape Cod, Washington, D.C., and Lake Champlain. And Christa slept easier knowing she could confide in him.

"He had faith in me," she said. "He made me strong."

When she was twelve years old, Teddy died and her father comforted her. A few days later, she heard him tell her mother about a silver cordial set he had seen at the jewelry store. He was disappointed, he said, because he wanted to buy it but it was too expensive:

twenty-nine dollars. Christa gathered all her babysitting money, bought the cordial set and surprised him with it the next day at dinner. She hugged him and told him she loved him.

Christa followed her father's advice and in the fall of 1961 entered Marian High School, a small Catholic school with a strong academic reputation and a strict code of discipline. The classes were taught by nuns who wore long black habits and starched white bibs with large crucifixes dangling from their necks. The girls wore traditional plaid skirts, white blouses and gray vests; the boys jackets, pants and ties. A giant St. Bernard roamed the halls with the Reverend William Shea, the school's spiritual director.

In many ways, Marian reflected the racial character of Framingham, particularly its Catholic community. A boy named Victor Chin was the only minority student in Christa's class, and most of the other 175 students had names that were either Italian — Ferraro, Gennaro and Zaccaro — or Irish — Murphy, O'Malley and Mulcahy. On the first day of Christa's sophomore year, the list of Irish names grew. A new boy arrived, a transfer student named Steven James McAuliffe.

Friends told Steve McAuliffe that the nuns

treated tardiness as if it were a mortal sin, so he arrived early to his homeroom, number 107, on the first day of school. A couple of seats remained empty when the bell rang, and he studied the teacher, Sister Seretina, a renowned disciplinarian, as she tapped her desk, waiting, her eyes fixed on the clock. Five minutes later, the door opened and a girl walked in. She was smiling. The nun narrowed her eyes. No one spoke.

Steve watched from his seat in the back of the room as the girl walked toward an empty desk, still smiling. As she sat down, he tapped the shoulder of the boy in front of him and asked in a stage whisper, "Who is that?"

"Christa Corrigan," the boy replied.

"She's beautiful," Steve said.

"Yeah, and *I'm* going out with her," growled a boy at the desk behind him.

Flustered, Christa smiled weakly, slumped in her seat and later told Anne Malavich that Steve McAuliffe was the cutest boy she had seen.

Homeroom was their only class together, so for a month Christa and Steve admired each other from a distance, trading uncomfortable smiles as they passed in the corridors or the cafeteria. No veteran at romance, Steve knew he wanted to introduce himself — he told himself so on the way to school every morning — but he had no idea how to go about it. He knew Malavich from his history class, however, and one Saturday morning Steve spotted Anne walking with Christa into a drugstore in downtown Framingham. Screwing up his courage, he followed them in.

The door had hardly jingled shut behind him when Steve began to feel weak in the knees, short of breath, suddenly very alone. Don't make a fool of yourself, he thought, his eyes flashing about the store. Just smile at her. That's enough for now. Don't push your luck. Don't take any chances. Just buy something and get out. . . .

Then Christa appeared, standing right in front of him at the candy counter. She was smiling. What could he do? He smiled back. What could he say? Nothing. His mind raced, but his lips were still. Seconds passed like hours, and then, mercifully, Anne introduced them. They were fifteen years old.

"I couldn't believe it," he said. "I was mortified."

"The rest is history," said Malavich, the maid of honor at their wedding six years later.

People at Marian High School remember Steve and Christa for their personalities as much as for their romance. Steve was a playful and aspiring student whose sophomore history teacher predicted he would become a lawyer because of his passion for debating such issues as the injustice of being born without one's prior approval. Christa was not the most beautiful or brilliant girl in her class, but she was one of the most spirited.

"There was a peculiar life to her, a special vibrancy," said Sister Mary Denisita, director of the school's community outreach program. "Her face was very alive, very interested. You could tell by looking at her that she was excited about everything life held before her."

Christa considered it a personal challenge when President Kennedy urged young people to give more than they took, to make a difference. She committed herself to charity programs in her church and her Girl Scout troop, and at Marian she helped start a holiday food-basket project for the needy.

"She never was just a joiner," Denisita said. "If she joined, she made sure she accomplished something. She really loved that kind of challenge."

And needed it.

"I've never felt good about myself if I wasn't giving," she said. "I felt like I was cheating myself and everyone else I had a chance to help."

The list of her other activities was one of the longest in her senior-class yearbook: glee club, student council, orchestra, basketball, ceramics, German club, the drama group. In her senior production of *The Sound of Music*, Christa played a nun, Sister Margaretta.

Marian was a college preparatory school, a breeding ground of competition, but Christa's classmates shared an unusual camaraderie. They feasted on the high-brow reputation they had among public school students, and when their football team, the Mustangs, faced one of the area's smaller public schools, they often joined voices in the Ivy League chant "Retard them, retard them, make them relinquish the ball."

Christa was a student of average intelligence who kept pace with her classmates and made the National Honor Society by studying with the kind of zeal that had made her a perfectionist in grammar school. By high school, she was simply a perfect overachiever. She graduated seventy-fifth in a class of 176, "but she was one of the few students you remember twenty years later," said Sister Lee Hogan, her social studies teacher. "She was always curious, always contributing."

Hogan's history courses, particularly her sections in current events, prompted Christa to consider a career in teaching. Before that, Christa knew only that she wanted to attend college, and it puzzled her that Steve had decided in eighth grade to be a lawyer.

"I couldn't imagine anybody making up their mind that early," she said. "I really admired him for that."

When she threw a surprise party for him on his sixteenth birthday, she gave him a block and gavel inscribed with "Steven J. McAuliffe, Esquire." She used the block and gavel years later when she served as a judge during mock trials in the law classes she taught.

Christa and Steve knew in their teenage hearts they were in love, and they dated nearly every weekend their sophomore year, pestering their fathers to drive them back and forth to St. George's Theater so they could hold hands and watch *The Great Escape, The Pink Panther, Shenandoah* — whatever was on the bill. Steve's father, Leo, encouraged them.

"Don't let that girl get away," he told his son.

But Steve's younger brother, Wayne, was confused. "Jeepers, Mom," he said, "I can see Steve going out with a girl for her looks, but not a girl with a brain."

Steve bought a motor scooter that year,

and, although Christa's parents forbade her to ride on it, he saw more of her than ever before. Christa's father grew edgy when they started talking about spending the rest of their lives together.

"Nothing against Steve," he told her. "He's a fine young man, but you just don't make up your mind about someone when you're fifteen. Why don't you see some other boys?"

Christa heeded his advice one night and dated a neighborhood friend. Her father waited up.

"How'd it go?" he asked.

"He was nice, and I had a good time," Christa replied, "but if I can't go out with Steve, I'd rather not go out with anyone."

"From then on," Ed said, "it was everlasting, Steve and Christa."

They dated during the school year and worked together in the summers — Christa as a swimming instructor and Steve as a lifeguard at Saxonville Beach on Lake Cochituate; Christa as a counter girl and Steve as a delivery driver for a dry cleaners. When Steve met Wayne Newton, Robert Goulet, Peter, Paul and Mary at two of the nightclubs on his route, he rushed back to the cleaners to tell Christa. And if they didn't see each other at night, they spoke on the phone.

"They only had two arguments the whole time," Grace said. "Christa would come home crying, but the next day everything would be back to normal."

One day when they were sixteen, Christa told Steve that if he asked her to marry him, she would say yes. He asked, she said yes and, to the relief of their parents and the nuns, they agreed to wait until they finished college.

But the engagement never kept Christa from her family. When she was fifteen, she made her father a ceramic manger scene that he lays out on a living-room table each Christmas. At sixteen, she made her mother a long red-velvet dress with white lace that she wears each year during the holidays. Christa taught her sisters how to swim and how to sew, and on Saturdays she took them on the bus when she went shopping.

"She never made us feel like we were tagging along," said her youngest sister Lisa. "We never felt like anything less because we were kids."

Christa slipped a souvenir — candy, a book, a party favor — under her brothers' and sisters' pillows whenever she came home from a date. And she checked on them when she came home late from babysitting, which was often. Christa babysat so much that her father remembered her "babysitting for the world," and on the July day in 1985 when she appeared on the news as the winner of NASA's teacher-in-space contest, a man sipping beer at a bar on Route 9 in Framingham leaped to his feet and shouted, "Hey, that's my babysitter!"

Of all her youth activities, Christa's best preparation for NASA's space contest was a Girl Scout competition she entered in the winter of her junior year in high school. For six months, she competed with girls around the world for a trip to the annual Girl Scout Roundup in Idaho. She filed applications and attended training sessions. She pitched tents blindfolded, gave emergency first aid and walked through unfamiliar neighborhoods to memorize details that would demonstrate her powers of observation. She endured dozens of tests and lectures to prove she could live on her own in the woods for a week. In July, she won a round-trip train ride to Idaho.

When Christa returned to Marian in the fall, she learned about the death of a classmate, Paul White, who had drowned during the summer in a swimming accident. She attended a memorial service for him and took to heart a poem that another classmate had written about mortality and the promise of heaven. The poem, "Somewhere," by William Dill, appeared on the last page of her yearbook:

We will meet again;
Somewhere we'll meet again,
Tall with reverence or bent in prayer.
Meet me there . . . somewhere.

Anytime at all
"Someone" may make His call.
Sorrow-laden we take our rest,
Crowned with glory to heaven blest,
Someday . . . somewhere.

Souls of the just glorifying,
Go where there's no pain or dying . . .
    somewhere.

We will meet again,
Our paths will cross again.
Reach for heaven with love and prayer,
Make a promise to meet me there
Somehow . . . someday . . . somewhere.

She learned at a class reunion fifteen years later that Dill had died of leukemia at the age of thirty-one.

Both of the colleges Christa applied to — Framingham State College and the University of Lowell — accepted her, but she passed up a chance to room with Malavich at Lowell so she could attend Framingham State and live at home. Her parents had urged her to go away to school.

"Save your money for the boys," she told them. "I'll stay here and get all the education I need."

The son of a career Army sergeant, Steve qualified for a scholarship at Virginia Military Institute, nearly six hundred miles away. He accepted it, and Malavich, figuring that distance would not pull apart the couple she had brought together, wrote in Christa's yearbook: "You're a real number one friend. I know you'll have every happiness, even though you'll have to put up with Steve. Good luck."

Steve wrote Christa a simpler message: "I Love You."

In Christa's final days at Marian, she lived up to Sister Denisita's image of her as a buoyant and devout student, but no goody-goody. Determined to wear what she wanted to wear to her senior prom, Christa defied the principal's order that no girl appear in a strapless gown. She arrived in long white gloves, a shiny bouffant hairdo and a lacy, full-length gown that left her shoulders bare. The nuns protested, but Christa stayed, bare shoulders, Steve and all.

In the fall, Steve went south to Virginia, and Christa joined her mother at Framingham State, Christa studying full-time for a degree in American history and secondary education, her mother part-time for an art degree. They bought a used Volkswagen bug and shared the five-minute ride to the campus.

Four months later, the nation mourned three astronauts — Virgil (Gus) Grissom, Edward White and Roger Chaffee — who died when a fire broke out in their spacecraft during a training session on the launch pad at Cape Kennedy. It was the darkest hour of the space race, but Christa, like all Americans, was reminded of what Grissom, one of the original seven astronauts, had once said about the people who explore the heavens: "If we die, we want people to accept it. We are in a risky business. . . . The conquest of space is worth the risk of life."

Commuter students often skirted campus life at Framingham State, but Christa carried a full load of courses, worked nights as a clerk at a trucking company, continued her baby-sitting binge, helped at home and still managed to captain the college debate team, sing in the glee club, perform in a college production of *The Pirates of Penzance* and help to organize a concert by Dionne Warwick. She made dean's list three times and rarely missed a class, even an early one.

"I was always a little crushed when it looked like Christa wouldn't make it, because I knew I could count on her to make a contribution," said Carolla Haglund, who taught an 8:00 A.M. course on the American frontier. "She always came through. She just had trouble finding a parking space sometimes."

Christa learned from Haglund the value of journals, historical accounts of the lives of ordinary people. She lay in bed at night and read the diaries of pioneer women like Susan Magoffin, who married at seventeen, the same day she set out in a covered wagon on the Santa Fe Trail, and of a frontier settler who wrote, "I taught the first school until a regular teacher could be found. I rode horseback and carried my baby on the saddle in front of me. I remember this period as the happiest of my life."

Haglund told her students that space was one of the last frontiers, and when she asked how many of them would like to go to the moon, Christa, seated in the front row as always, raised her hand. Neil Armstrong was far ahead of her, however.

In December 1968, astronauts Frank Borman, James Lovell and William Anders circled the moon ten times in history's first manned flight from Earth to another body in the solar system. As they soared through space on Christmas Eve, they read the opening passage of the Bible to millions huddled before televisions back home.

"In the beginning," Borman said, "God created the heavens and the Earth . . . "

Seven months later, at the historic moment Armstrong stepped onto the moon's Sea of Tranquillity — 10:56 P.M. on July 20, 1969 — Christa and Steve were driving through a rainstorm in Pennsylvania. The highway hissed beneath them and their windshield wipers beat like a tin soldier's drum, but the radio was loud enough so they could hear Armstrong's voice crackle from his perch 230,000 miles above the clouds.

"That's one small step for a man," he said, "one giant leap for mankind."

They cheered, and Christa's stomach tingled. When Eastern Airlines offered reservations on the first commercial flight to the moon as a public relations gimmick, she urged Steve to sign up with her. And two months after the moon landing, she marveled at Vice President Spiro Agnew's prediction that an American — not necessarily a man — would step foot on Mars in 1986.

Mars? The idea fascinated her, but in the fall of 1969 Christa was more concerned about setting foot on Steve's college campus. She had a social life of her own, skiing in New Hampshire, crashing college parties with her friends from Framingham State, attending a rock concert by the Jefferson Airplane in Boston. But she dated no one but Steve, and the weeks between trips to VMI often seemed like months.

A light-hearted rebel, Steve complicated Christa's first few visits by committing a series of minor offenses that helped him set a record for making and losing rank. Twice he was promoted to corporal and sergeant, and twice he lost his stripes. He was elected vice-president of the Malcontent Club, and despite his reputation as the barracks lawyer, he failed to talk his way out of walking penalty tours every Wednesday and Saturday until April of his sophomore year. When Christa visited, she spent her Saturdays walking with him, back and forth, back and forth, hours at a time, smiling.

"I don't know what the rules were, but

Steve didn't mind," said Christa's college friend Joanne Brown, who usually went along. "I know Christa sure didn't."

Washington was on the road to VMI, and Christa rarely missed an opportunity to stop there on the way home. She toured the White House, visited the Smithsonian, sat in on Supreme Court hearings, discovered something new each trip. The city intrigued her.

But, like Washington, Christa changed in her years at Framingham State. The war dragged on in Vietnam, claiming as many as two thousand American lives a month. Four students were shot dead by the National Guard during a protest demonstration at Kent State University. Martin Luther King, Jr., was assassinated in Memphis; Robert Kennedy, her idol's brother, was gunned down in Los Angeles; and Mary Jo Kopechne, Robert's former secretary, drowned in a car that plunged off a wooden bridge on Chappaquiddick Island off Martha's Vineyard. The driver was the youngest Kennedy brother, Ted.

"I became aware of an awful lot in those years," Christa said. "I developed a healthy distrust for authority. I guess I started hoping I could change the world."

She bore the spirit of Ralph Chaplin's poem "Mourn Not the Dead," which appeared in her college yearbook:

Mourn not the dead that in the cool earth
   lie —
Dust unto dust —
The calm sweet earth that mothers all
   who die
As all men must;

Mourn not your captured comrades who
   must dwell —
Too strong to strive —
Each in his steel-bound coffin of a cell,
Buried alive;

But rather mourn the apathetic throng —
The cowed and the meek —
Who see the world's great anguish and its
   wrong
And dare not speak!

Christa protested against the war, wore a black armband to her graduation and spoke louder than ever about the rights of women, blacks, native Americans — anyone she believed deserved a better deal. She wore the uniform of the time: a bleached Afro hairdo, horn-rimmed glasses, white lipstick and billowing paisley dresses. She called it her "rad-

73

ical period," but she avoided two common curses of the period — extremism and escapism.

"She never strayed too far to the left or the right, and I never worried about her getting into drugs or anything like that," her mother said. "She had everything in perspective."

After all, she was a Girl Scout, the leader of a troop in which her sisters were among the scouts she took sightseeing in Washington and skiing in New Hampshire. And she was still daddy's little girl. When Steve was unable to attend her senior dinner dance at Framingham because he was stuck in Virginia on maneuvers, Christa invited her father.

"Why?" her father wanted to know.

"First, because Steve can't be here," she said. "Second, because you're a better dancer than he is, and third, because I have enough respect for myself that I won't feel uncomfortable going with my father."

So they went, their last time together — just the two of them — before Christa and Steve were married. That spring, seven years after Christa walked in late to Sister Seretina's homeroom, Steve's college yearbook editors predicted "his future — Christa at last."

CHAPTER THREE

At last they were married, eight weeks after they graduated from college and two days before Steve entered the Georgetown University Law School. It was a drizzly Sunday, August 23, 1970, and Christa walked down the aisle of St. Jeremiah's, the family church, with daisies in her hair and her hand on her father's arm. Steve waited at the altar, his chin thrust forward above a black bow tie, his only brother, Wayne, by his side. A guitarist played the love theme from Zeffirelli's *Romeo and Juliet*.

After the ceremony a local rock band, The Trolls, performed under a yellow-and-white tent in the Corrigans' backyard. More than two hundred people celebrated by dancing and drinking until the newlyweds drove away at dusk with everything they owned stuffed into a rented trailer that swung from the back of their aging, orange VW bug, tin cans dragging behind it. They were twenty-one years old, jobless, and had five hundred dollars between them. They honeymooned on the highway.

"We thought we had everything we needed," Steve said. "We never figured Christa would have trouble finding a job a week before school started."

Not until the Washington skyline came into sight did it strike them that teaching vacancies, especially in social studies, were as likely in late August as ice on the Potomac River. But Christa, charmed as ever, found work within a week as a full-time substitute teacher at Benjamin Foulois Junior High School in suburban Morningside, Maryland. She taught American history to eighth graders for a year, then spent the next seven years teaching English, American history and civics at Thomas Johnson Junior High, a troubled, primarily black school in a lower-middle-class neighborhood of Lanham, Maryland, fifteen miles from the capital.

Her salary helped pay the rent on a tiny tenement apartment, which came with a leaky roof and roaches that were "bigger than the furniture," in a depressed section of Washington. Christa and Steve supplied the furniture: inflatable zebra-striped sofa chairs; orange-crate tables; lamps and curtains that suggested a twentieth-century flea market motif.

"We had everything we needed," she said.

They settled into a more comfortable apartment in 1972, and Christa brought home their first cat, Rizzo, which she named for two of the less glamorous characters of the time, the reactionary mayor of Philadelphia, Frank Rizzo, and the pathetic hero of the movie *Midnight Cowboy*, Ratso Rizzo.

When she arrived at Johnson Junior High that fall, "it was considered one of the more difficult schools in the state to work at," said Thomas Campion, a social studies teacher who started the same time. But Christa soon gained a reputation as the teacher most willing to work with problem students. An overcrowding crisis magnified the challenge by forcing her to use a corner of the library as a classroom.

Undaunted, Christa wheeled in a temporary blackboard, decorated the walls with *Time* magazine covers and started each class with the "name game," weaving through the aisles and asking the students to repeat their names until everyone in the class knew every classmate's name. She wrote her own name on the blackboard and told her students that she would deduct points each time they spelled it wrong.

Sometimes she kept them interested by taking risks that amused her colleagues. One day she walked into school with a guitar slung over shoulder and, standing before a class of history students whose musical tastes began and ended with the Top 40 tunes of the time, Christa sang folk songs they had never heard, songs by white artists such as Pete Seeger and Woody Guthrie. They gawked at first, but by the end of the class they joined voices and the sound of "This Land Is Your Land" echoed through the building.

"Another teacher might have thrown up her hands in disgust," said Campion, now the social studies department chairman. "There were plenty of distractions, God knows, but Christa went the extra mile and made it work."

She enjoyed the guitar, and when colleague Patricia Mangum asked her for lessons, Christa threw herself into the assignment. They sat in Mangum's living room, Mangum strumming and Christa singing her favorite tunes: "Blowin' in the Wind," "Amazing Grace," "The Times They Are A-Changing," "Puff the Magic Dragon," "Leaving on a Jet Plane," "Man of Constant Sorrow."

She also developed her favorite teaching tool — the field trip — in her years at Thomas

Johnson. Christa's history students traveled to Williamsburg, Jamestown, Gettysburg and historic places in and around the capital. Her law students went to prisons and courts, sometimes to see Steve try a case. She encouraged all her students to learn as much from seeing and doing as they could from reading and listening, unaware that in the winter of 1973 some of them would learn a history lesson the hard way.

Dozens of students at Thomas Johnson, like thousands across the country, were uprooted in mid-year and transferred to unfamiliar, mostly white schools under court-ordered desegregation. To soften their anxiety and to help keep homework and other important papers in order during the first days at their new schools, Christa gave the students folders with their names on them. She tucked a personal note of encouragement into each folder and visited each of their new teachers to try to ease the transition.

Two years later, Christa entered a new school of her own. Convinced she could make a greater mark on education if she was a policy maker, Christa began a master's program in education supervision and administration at Bowie State College, a school that, like Framingham State, was convenient, inexpensive and small enough so professors could work closely with their students. Campion sat next to her in a summer course on education law.

"She was extremely determined, especially in that law class," he said. "I never was sure if it was because she wanted to prove herself to her husband or because she just wanted to do the best she could do at everything she tried. Whatever it was, she got an A."

She soon practiced her college lessons. Christa was the chairperson (she insisted on "chairperson") of the faculty advisory committee at Thomas Johnson, and when the principal, Thomas Perry, left for several months to undergo a kidney transplant, he asked the committee to replace him. Christa "pretty much ran the school," said Mangum, the committee's secretary. "She was a strong leader, very outspoken."

Christa also entered union politics, describing herself as a Kennedy Democrat, a feminist whose "sympathies have always been for working-class people." Dissatisfied with the Prince George's County Education Association, the bargaining agent for the school's teachers, she waged an unsuccessful campaign to elect the American Federation of Teachers as the new bargaining agent.

"She didn't pull it off, but she definitely

gave it her best," Mangum said. "She loved the give and take of politics."

In eight years of Washington life, the McAuliffes rose from struggling graduates to promising professionals. Steve earned his law degree, clerked in the private practice of Steny Hoyer, the president of the Maryland State Senate, and met his military obligation by defending court-martialed servicemen for four years in the appellate division of the Army Judge Advocate General's Corps. Christa taught full-time, waitressed part-time at a Howard Johnson's and earned her master's degree after completing a thesis on "The Acceptance of the Handicapped Child in a Regular Classroom by His Normal Peers."

Their careers on course, Christa and Steve bought a modest split-level home in a racially mixed neighborhood of Oxon Hill, just north of the city. A few months later, they attended the Metropolitan Opera with Christa's great-great-aunt Carrie, a ninety-year-old matron who lived alone in the Spanish Harlem section of Manhattan.

Carrie had entertained Christa's family often through the years, and at the age of seventy-six she had treated all seven of the Corrigans to a week at the 1964 New York World's Fair. Later she had invited each of the children to attend the Met with her and dine at Luchow's, a pricy Manhattan restaurant. On this visit, Christa and Steve learned that developers planned to turn her apartment building into condominiums. Worse, she was losing her sight.

They returned to New York with a rented truck the following weekend. They packed up Carrie's belongings and took her home with them to Maryland, where she lived with them for a year, a quick-witted woman of German ancestry who drank a glass of beer each day with lunch and sipped a little whiskey each night before bed. At dinner she'd talk about the opera — she had attended Enrico Caruso's first appearance at the Met in 1903 — and tease Steve about his Irish roots. Afterward, she often played the piano and sang.

In the mornings, Carrie played a little trick on the McAuliffes. She liked the house warm, very warm, but Steve and Christa preferred it cool. So after they went to work each day she cranked up the thermostat, always sure to turn it down before they returned. They often wondered why their home felt like a bathhouse, and they found out one day when they arrived unexpectedly for lunch. The temperature was 90 degrees and Carrie, stunned by their sudden appearance, was scurrying toward the thermostat.

"Aunt Carrie," Steve said, "are you all right?"

"I'm fine," she huffed, pink with embarrassment. "But how many times do I have to tell you to keep the heat down in here."

Carrie's health was failing, but she went to church each Sunday with the McAuliffes and stayed up late to attend their election-night parties and Halloween costume balls, often using her poor vision as a source of humor. She answered the door at one Halloween party to discover a guest dressed like Dracula.

"Oh, dear," she said, closing the door in his face. "I don't think Steve and Christa would like to see you here without a costume."

On a Sunday morning soon after that Halloween, Steve and Christa were ready to take Carrie to church when she told them she was tired and needed to rest a bit. When Christa checked on her fifteen minutes later, she was dead.

"That year with Carrie was one of the best of our lives," Steve said. "She gave us a laugh a day, but more than that, she was a wonderful role model."

Like the pioneer women she studied at Framingham, Christa began to keep a diary after Carrie's death. In a spiral notebook she recorded everything from her deepest personal feelings to the antics of the family cats. On February 1, 1976, she learned she was pregnant with her first child. Christa was twenty-seven.

"How appropriate for a social studies teacher to have a bicentennial baby," she wrote. "The D.C. area is a flurry of activity . . ."

She kept the journal to communicate the experience of a young mother to her child, to share the lessons of one generation with the next.

"When the children are old and they're starting to think about their own careers and marriages, I'll be able to hand them part of my life," she said. "I would have loved it if my mother had said, 'Here: this is what my life was like.'"

On March 11, 1976, she wrote, "I felt like I should do something, so I got some books out of the library and read all of them. I'm still so confused."

She detested milk, so she ate yogurt as a substitute during her pregnancy. She ate brown rice and granola and drank Constant Comment tea long before they were fashionable, and she continued to enjoy her time in the kitchen, testing recipes she collected from friends, relatives, magazines, newspapers, even chain letters. She never shied from a culinary risk. For a dinner party one night, Christa tried a recipe for Beef Wellington that her father had sent her. Her guests raved about it.

At 5:00 A.M. on Sunday, September 11, 1976, Christa gave birth to Scott Corrigan McAuliffe in the hospital at Andrews Air Force Base. Steve later called it the happiest day of his life, but at the time he was more worried about missing his favorite football team, the Washington Redskins, on television that afternoon. As Christa cradled the new baby in her hospital bed, Steve sat by her side and watched the game. He wrote in Scott's baby book, "Be proud that on your first day on this Earth you watched the Redskins beat the Chicago Bears."

Two years later, Steve completed his Army stint and the McAuliffes came to a crossroads. The U.S. Justice Department offered him a staff position, a tempting opportunity that Christa gently reminded him was not part of their plan. They had agreed to raise their family in New Hampshire.

Christa had summered in New Hampshire as a Girl Scout and skiied there in the winter.

She liked the state, its beauty, its easy pace and now, as a practical matter, its distinction as the only state in the nation without a sales or income tax. Steve argued his case on professional merits. Christa could be a lawyer, he said, a good one, and they could have wonderful careers right there in Washington. He even talked her into taking the law boards, but she would go no further.

"She realized she would make a much greater impact as a teacher," he said. "She knew she would impart knowledge and life values to thousands of children before she was through."

Steve's case continued to crumble when they visited an old Army buddy, Terry Shumaker, and his wife, Polly, in Bow, New Hampshire, an attractive bedroom community next to Concord, at Christmastime in 1977. Christa "fell in love with the state all over again," Steve said, and, sensing defeat, he did what any reasonable lawyer would do. "I made a deal with her."

The deal was that Steve would apply for a job as an assistant New Hampshire attorney general in Concord, the state capital. If he got the job, the family would go north. Otherwise, they would stay in Washington. Christa considered the deal, balked, and offered another suggestion.

"You can live where you want to live," she said. "Scott and I will live in New Hampshire."

The attorney general settled the matter by hiring Steve as an assistant in the civil division. Christa applied for work as a substitute teacher and the family moved into an aging brown-shingled Victorian on the "Hill," a neatly groomed Concord neighborhood where they lived across a small park from the city's newspaper publisher. The house needed a complete renovation, but Christa was eager to start. They moved in during the blizzard of 1978, nearly twenty inches of fresh snow in their driveway.

Concord was a city in the country. Strawberry fields flourished in summer two miles from the state house, its golden dome glistening on a squat skyline. A turkey farm produced Thanksgiving birds for most of the city's citizens and stocked the local supermarket shelves with enough frozen pot pies to provide a lasting alternative to the mile-long neon strip of fast-food restaurants known as Hamburger Hill. Cornfields hugged one bank of the Merrimack River as it twisted through the city on its 120-mile jaunt to the Atlantic; shopping malls and car dealerships bustled on the other bank.

Pronounced "Concuhd" by Christa and New Englanders — "Concorde" by everyone else — it was a family town of thirty thousand that Christa described as "a Norman Rockwell kind of place." The crime rate was low, the schools were good and recreation facilities abounded. Drivers stopped politely for pedestrians on Main Street, and check-out lines at the supermarket often turned into social gatherings. Parents spent their weekends ferrying children between sports practices, puppet shows and church. The nightlife, but for a few restaurants that closed early, was just about how *The New Yorker* described it the year the McAuliffes arrived: "By and large, Concord is a government town, with quiet state employees. . . . Andrew Jackson could prance down Main Street on his horse some Saturday night and the chances are nobody would see him. Nobody out."

Many families would be home after canoeing on the Contoocook River, casting for bass at Turkey Pond or hiking through the city's wildlife preserve. If it was winter, they might have skated on the frozen duck pond at White Park, cross-country skiied on the grounds of the Audubon Society or built snowmen on the state house lawn. The children would be in bed; their parents would be sitting before televisions.

Like the McAuliffes, most people in Concord lived comfortably. Condominiums were occupied as soon as they opened, people stood in line to buy crabmeat quiche at a new gourmet shop and tickets to hear Itzhak Perlman at the Capitol Theatre were gone within hours. But amid the prosperity lay pockets of poverty: weather-beaten shacks beside a rusting railroad line, mobile homes with two children to a bed, rundown tenements within blocks of the capitol's gleaming dome.

Christa soon learned about her neighbors at the bottom of the Hill. Her first teaching assignment was to tutor an elementary school student who had been suspended for assaulting a custodian. School district officials feared Christa might not be safe in the student's home, so they asked her to meet with him daily at the public library, a block from the police station. Mark Beauvais, the superintendent of schools, monitored their progress.

"I was amazed at her ability to keep him interested," he said. "She knew just when to push or pull back, and worked with him very successfully under the worst kind of circumstances."

The boy improved enough to return to school, but he has since turned twenty-one and landed in jail.

"We didn't save him," Beauvais said, "but Christa did all she could."

When a teaching vacancy opened at Rundlett Junior High School the next fall, Beauvais pounced, hiring Christa to teach English and history, although budget cuts forced him to lay her off at the end of the school year under the last hired, first fired policy. She was seven months' pregnant.

In six years, Christa had more children than America had people in space. The Soviets ruled the heavens while NASA dedicated itself to building a space-age covered wagon, a winged glider whose giant cargo bay would transport the tools and equipment needed to colonize the high frontier. It would be the world's first reusable spacecraft, and its passengers would include Francis Scobee, Judith Resnik, Ellison Onizuka and Ronald McNair, four of the thirty-five astronauts who were selected from a crush of more than eight thousand applicants in January 1978. They were a military pilot, a Jewish woman, an Asian American and a black — symbols of NASA's commitment to carry America's cultural rainbow toward the stars. If they succeeded, Christa thought, space cities one day would thrive with ordinary citizens, maybe her children.

On August 24, 1979, the day after their ninth wedding anniversary, Steve drove Christa a half mile down a bumpy road to Concord Hospital, where she delivered their daughter, Caroline Corrigan McAuliffe. They gave the baby the same name as John Kennedy's first daughter and bought her a gold cross and chain they hoped she would pass on to their grandchildren. They had a boy and a girl. Their family was complete.

Restless, Christa returned to the blackboard two months later, teaching English and social studies full-time to ninth graders at the Bow Memorial School. She continued to take her students out into the world and increasingly brought in new faces: policemen, hairdressers, body builders, auto salesmen — ordinary people who could tell young people what life was like beyond the schoolyard. As always, Christa allowed her students to talk quietly in class (a practice other teachers warned her against) because she believed they needed to feel comfortable when it was time for an organized discussion.

"We have a classroom based on mutual respect," she told her students at the start of each school year. "I ask only two things of you: be yourself and do the best you can."

In her spare time Christa jogged, played

tennis and volleyball, skiied, sewed and phoned old friends late at night to sing Girl Scout songs.

One night in mid-stanza she had an idea. She arranged for five friends from her old Framingham Brownie troop to get together for a weekend at a Girl Scout camp they had attended fifteen years earlier in the woods of Bradford, New Hampshire, thirty miles from Concord. Each of the women had a career, a husband and at least two children, but drawn by memories and friendship they packed knapsacks and drove off to pitch their tents on a bed of pine needles by a mountain lake. Christa's mother, their former troop leader, also joined them.

For two days they hiked, canoed, fished and forgot about finding lost socks for their children and cleaning up after their husbands. At night they sang by the campfire and shared stories of life's passages, moonlight glittering on the lake. They returned to the camp every summer after that.

At their reunion in 1983, Christa told them about the death of her grandmother. Christa had been the daughter her grandmother had never had. They had been close through the years, companions and confidantes, and Christa's fondest childhood memory was a

Christmas morning in 1952 when she sat in a little red rocking chair beside a freshly decorated tree in her grandmother's living room. She was smiling at her grandmother and her grandmother was smiling back, a simple joy that lasted a lifetime.

When her grandmother died, Christa asked her cousin, the Reverend James Leary, if she could read from the Bible at the funeral mass. Her chin trembled as she began to read, her eyes welled with tears and the words stuck in her throat. Sobbing once, Christa raised her finger to her lips to steel herself, pausing before she went on. Two years later, she calmed herself the same way at the White House.

Back at Bow, Christa asked her students to keep diaries, and on April 12, 1981, exactly twenty years after Yuri Gagarin's historic mission, she made sure they took note when John Young and Robert Crippen christened a new age of space flight by circling the Earth for fifty-four hours in the shuttle *Columbia,* a flying machine that landed on a desert in California as smoothly as a corporate jet could land at the Concord airport.

"We've got a fantastic and remarkable capability here," Young said upon his return. "We're really not too far — the human race —

from going to the stars."

It sounded wonderful to Christa, but first she faced a more earthly challenge. As president of the Bow teachers union, a chapter of the National Education Association, she faced a town of uneasy taxpayers who felt squeezed by a state government that ranked forty-ninth in the nation in funding public education. The school board intended to hold down the budget by capping teachers' salaries, which started at less than $10,000. The contract was about to end, and Christa was about to negotiate.

If she was concerned about her image, about providing ammunition for people on the other side of the bargaining table, she hardly showed it in her morning routine. Teachers were required to arrive at the Bow school by 7:30 A.M., nearly an hour before the first bell rang; but Christa, with two children and a husband at home, usually arrived just before the bell, her hair still wet, her students waiting.

"It wasn't too cool," said a close friend and colleague, Eileen O'Hara, "but I don't think anybody said anything."

In fact, Christa did most of the talking, including a speech at a school district meeting in which she warned of the perils of underpaid teachers and inadequate education. She convinced the townspeople to approve a bigger budget and increase teachers' salaries. But a year later Christa was gone, angry that she had been passed over for a job as the assistant principal at Bow because the administration "wasn't ready for a woman yet."

The superintendent in Concord had offered her a teaching job at the high school under the last hired, first fired policy. She had accepted.

Concord High School was three blocks from her home and, better yet, it was well funded, progressive and regarded as one of the best public schools in the state. The worst thing an independent team of educators could find to say about the school the year Christa arrived was that the cafeteria needed more supervision, a charge the principal, Charles Foley, denied vehemently. On the day of the evaluation, he argued, sixty members of a visiting Belgian boys choir, some with a penchant for tossing peanuts, had disrupted the lunch hour.

"We haven't had a food fight in nine years," he protested.

Flanked by convenience stores a block from the state mental hospital, Concord High rambles in red brick through a residential neighborhood. Giant stone pillars rise from

95

the front steps of the main building, a refurbished relic of the 1920s, where Christa
taught in a well-lighted room behind a large
wooden door on the third floor. Her room,
number 305, was adorned with magazine covers and newspaper clippings that summarized
the public issues of the times: abortion,
women's rights, the militarization of space.
On their way into the building, her students
passed a stone statue that quoted from Shakespeare's *Julius Caesar:* "I love the name of
honor more than I fear death."

Christa taught American history, law and
economics, more intent than ever on making
her subjects come alive in the classroom. In
history, she divided her class into groups and
assigned each group a decade to re-create in
class. They dressed in period clothes, performed the songs and dances of the time and
acted out monologues reflecting the prevail-
ing political attitudes. Her economics students
talked to stockbrokers, bought imaginary
stock and followed it in the newspapers. Her
students in "You and the Law" examined the
legal issues of the day, staging mock trials to
study the rights of landlords and tenants, em-
ployers and workers, juveniles and the police.
They called the course "Street Law."

Teaching was Christa's challenge, her daily

exercise in living on the edge.

"Every forty-eight minutes I get a new group of kids," she said. "I can't sit back and think about it for fifteen minutes. They're going to come at me whether I've had my coffee in the morning, whether I have a headache, whatever the problem might be. That's what makes it so exciting. One day is never like another."

She had fun with it.

"True or false," she wrote once in an exam, "Tom Herbert [her department head] looks like Trigger."

It was a touch of comic relief, her way of soothing anxious students. The next day Herbert burst into her law class, his face flushed, his jaw taut. "What you did to me was a flagrant act of insubordination!" he said, spitting out his words.

"Don't you dare barge into my class like that," Christa countered, and the exchange continued until Herbert threatened to have her fired and whirled away.

"My God, I'm devastated," she said, slumping at her desk, resting her face in her hands. "Everyone, please, write down what you just saw. I need your help."

The drama, of course, was just that — a lesson on the role of being an eyewitness. Herbert, who had known in advance about the question, later returned to the class to hear each student's impressions of the encounter.

Not every day was so tumultuous, however. Looking very collegiate in a gray wool vest and a tweed skirt on an April morning in 1985, Christa began her law class by discussing articles in the previous day's newspaper. She read aloud about Gary Dotson, who served six years in an Illinois prison before the woman he was convicted of raping recanted her story. Now he was asking the governor for a pardon.

"Didn't Ford pardon Nixon?" a boy asked.

"Ford pardoned Nixon for the crimes he *might* have committed while in office," Christa said. "Isn't that wonderful? Wouldn't it be wonderful if someone pardoned you for everything you *might* do wrong?"

Then she turned to a Supreme Court case that ended in the banning of prayer in public schools. The case began when William Murray, who was raised an atheist, came home from school one day in 1960 and complained to his mother that his teacher made him say the Lord's Prayer each morning.

"He was challenging his mother," Christa told the class. "He was doing what a lot of fourteen-, fifteen- and sixteen-year-olds do.

He was saying, 'We don't believe in God, and I have to pray in school. If you don't do something about it, you're a hypocrite.' "

The class shifted from explanation to discussion to performance, and, like the nuns she had learned from, Christa allowed no nonsense. She was strict on behavior but free on ideas, urging students to share their opinions, reminding them that people with different values understand events differently. She asked them to see the case through William Murray's eyes.

"If students can see how an event may impact on them, they remember it better," she said after the bell rang. "Like Supreme Court decisions. Who thinks about Supreme Court decisions? But if you relate it to a young boy who is having a problem in school, it makes them think about it."

Christa was a social historian, a student of "all those good people who lived and worked in our history and you never hear about." Ordinary citizens were as important to the study of history as presidents and kings, she believed, and in her second year at Concord High, she merged that philosophy with her feminist values to develop a course on the American woman. Her sources ranged from eighteenth-century pioneer journals to *Her-*

story, a woman's perspective of American history by June Sochen. She had planned to call the course Herstory, but then decided it sounded too radical.

"The American Woman" was not the most popular course at Concord High. Only fifteen girls and one boy signed up, but Christa pressed on, starting each semester with a lesson on the American Revolution — not the battles between men at Bunker Hill, but the struggles of the women at home. While the men fought, she told her students, women fed their families, sewed their uniforms and brewed their liquor. Women contributed in many ways to help win the war.

"Is that an unusual role for women — to donate things?" she said. "No, that's a really traditional role for women, even today, helping out a family in the neighborhood, for instance. . . . Women as a whole will become involved with a neighborhood a lot better than men because of the children."

Christa was not recognized as the best teacher on a staff of ninety at Concord High. In fact, three of her colleagues, Herbert among them, were named New Hampshire teachers of the year between 1978 and 1985. She shared their enthusiasm, though, their conviction that every day provided a chance, a new opportunity to improve a child's mind, to make a difference.

She told her students about her early days in Washington, the roaches and the leaky roof, and explained that fifteen years later she and her husband were doing exactly what they wanted to do.

"Any dream can come true if you have the courage to work at it," she said. "I would never say, 'Well, you're only a C student in English, so you'll never be a poet.' You have to dream. We all have to dream. Dreaming's okay."

After sitting in on one of her classes, the principal, Charles Foley, wrote in Christa's evaluation: "It's reassuring for me to see a first-class teacher at work." He had no suggestions, no criticisms. She was a master teacher.

Her colleagues had only two complaints: her obsession with field trips that too often made her students miss their next class, and her annoying habit of putting everything off to the last minute, of rushing in late to staff meetings with Caroline in her arms. She was disorganized, some of them said, overcommitted.

"She loved her job, but there were so many other things, too," said Malavich, who, like

Christa, became a high school teacher. "Any woman who has a full-time job, a family and a home to take care of is naturally busy. Christa extended that through the other areas of her life, and it finally got to the point where she just had to be busy."

While Christa stayed busy, Steve represented the state in legal cases that ranged from construction of a nuclear power plant on the New Hampshire seacoast to a governor's order to lower the state's flags to halfmast on Good Friday in commemoration of the death of Christ. Steve's modest nature and ordinary appearance — he was slightly pigeon-toed, a few pounds too heavy and beginning to lose his hair — sometimes helped him in court as much as his keen legal sense.

In the flag-lowering case he defended the governor, Meldrim Thomson, against five clergymen who considered the order a violation of the constitutional separation of church and state. The case went overnight to the U.S. Supreme Court, where Steve and his partner, Wilbur Glahn, tried to file briefs they had scribbled on legal pads on the flight to Washington. The clerk refused to accept them.

Stunned, Glahn stood in the hall and wondered how he would explain it to the governor. Steve returned to the clerk's office.

"It's all set," Steve said when he came back. "The clerk changed his mind."

"How'd you do it?" Glahn said.

"Easy," Steve said. "I told him if he didn't accept them, I'd appeal to a higher court."

The case ended with both sides claiming victory.

In 1982, the year Christa went to Concord High and four years after Steve joined the attorney general's office, he entered private practice with Terry Shumaker, his old Army buddy, as a trial lawyer for the Concord firm of Gallagher, Callahan and Gartrell. His career was secure, both the children were in school and Christa had more time to pursue a lingering passion — community activity.

She led a Girl Scout troop, taught catechism to eighth graders at St. Peter's Church and volunteered as a receptionist at a family planning clinic. She joined fund-raising campaigns for the YMCA and the hospital. She powdered her hair and played Granny Greenthumb in a community theater production of *Little Red Riding Hood and the Monkey Flower Wolf*. She marched down Main Street in costume parades on New Year's Eve and performed a comical version of "Leader of the Pack" in a faculty talent show.

For exercise, Christa played volleyball with

the Court Jesters in a competitive co-ed league. She formed a tennis foursome at the Racquet Club of Concord, and jogged enough that she could complete the Bonne Belle 10-Kilometer Road Race — in which seventy-five hundred women jostled through the streets of Boston — in a respectable time of forty-eight minutes, good enough to finish among the first thousand.

"Christa, how can you do all this and keep it straight?" her father asked.

"All I'm doing is following my mother," she replied. "If she could do it, so can I."

Her daily appointment lists helped. Christa slipped one under a magnet on her refrigerator, taped another to the dashboard of her Volkswagen bus and shoved a third one in her pocket. The lists often started with laundry before breakfast and ended with more laundry before bed, but they seldom took priority over her students: a former junior high student who stopped by each day to park his bike in her garage; an inner-city exchange student who needed a friend and a home-cooked meal; a group of students who turned up on her doorstep on weekend nights to share their pizza and problems; a girl who called at midnight to say she wanted to die.

When the girl mentioned suicide, Christa

convinced her to come by the house to talk. They sat on a couch in the family room for a while and then Christa stayed up all night as the girl slept in the bunk bed below Scott's. When Steve later warned her about the potential legal problems, she had two words for him: "Sue me."

Christa's school days were harried, but her summers were her own. She painted and wallpapered the house, weeded a small garden and kept the bird feeder full on the small crabapple in the front yard. She took the children to feed the ducks at a nearby pond, and read her favorite Robert Ludlum spy novels as they played by the pool at the Concord Country Club. She waited for the annual Girl Scout reunion and prepared for a Labor Day party she hosted each year to celebrate her anniversary, her birthday and the children's birthdays, which fell within three weeks of each other.

On August 27, 1984, just after she returned from the reunion and a couple of days before the party, Christa and Steve were driving through Concord when they heard an interesting news item on the radio, something about private citizens in space. They turned up the volume.

"Today," the president said, "I am directing NASA to begin a search in all of our elementary and secondary schools, and to choose as the first citizen passenger in the history of our space program one of America's finest: a teacher."

Teachers with five years of experience could apply, the announcer said. Christa felt her stomach start to tingle. She smiled at Steve, and he smiled back.

"It's a don't miss," he said. "Go for it."

CHAPTER FOUR

Long before the president announced a teacher would enter space, ordinary men and women wanted to go. Thousands asked NASA for a lift into orbit after Alan Shepard snapped his heels together, saluted his Redstone rocket and rode it into history. Their spirit soared with the moon landings, tailspinned in the seventies and rose again in 1981 when John Young and Robert Crippen, dining on shrimp cocktail and peach ambrosia, christened a new era of space travel with the maiden flight of a reusable shuttle. Space suddenly seemed as accessible as a shopping mall.

Astronauts no longer wore bulky suits, no big-bubbled helmets, and the brutal force of gravity that grabbed Shepard on lift-off now was no fiercer than a carnival ride. The cubes-and-tubes diet John Glenn endured as the first American to orbit the Earth had been re-placed by a menu that featured beef strogan-off, shrimp creole and turkey tetrazzini. The shuttle's bathroom accommodated both sexes, and, like ordinary tourists, space travelers of the eighties packed cameras, binoculars and stereo cassette players. The ride seemed so routine that Rockwell International, the shuttle's manufacturer, blueprinted an orbiter that would accommodate seventy-four passengers, a sort of twenty-first-century tour bus.

"In 1985, we are in the very beginning of emigrating from our planet," said Scott Carpenter, one of the original seven astronauts. "We are beginning an exodus."

The first outpost was to be a space station, an orbiting city, which NASA expected to complete by 1992, five hundred years after Columbus dropped anchor down range from Cape Canaveral. Shuttle riders would help build the city, and ordinary people wanted to join them. A real estate developer in New Jersey offered $50,000 for a round-trip ticket to space. A Detroit man wanted to sacrifice his appendix to undergo the first orbital surgery. A barber proposed to give free "moon cuts" aboard the shuttle. Lawyers wanted to help NASA's legal department determine exactly where the Earth's atmosphere ended. Reporters craved the ultimate dateline: SPACE. A few people hoped the shuttle would carry them closer to God. An eighty-six-year-old man wrote that he jogged ten miles a day, was perfectly healthy and deserved to represent the elderly on the shuttle. Children complained that dogs, bugs and congressmen were flying, but no children. A singer, John Denver, pressed NASA to let him compose a tune in orbit. Couples wanted to consummate their marriages on the shuttle. Women wanted to deliver the first baby in space.

More than ten thousand pieces of mail a year — letters, checks, videotapes, proposals from police officers, poets and painters — poured in, and NASA, more dependent than ever on public support, began to respond. In 1982, NASA's administrator, James Beggs, asked author James Michener, astronaut Richard Truly and five others to explore "the difficult problems of selecting from a very large list of private citizens those who will be permitted to fly on the shuttle." A year later, they issued a "cautious go-ahead."

"It is feasible for NASA to fly [private] individuals on the shuttle beginning in the mid- to late '80s," they wrote to Beggs. They warned, however, that people may "easily misunderstand such a program as a self-serving public relations gimmick that trivializes the space program." Also, "there are questions of flight risk, liability and the control of extraordinary profits" involving citizen passengers, they said.

Their conclusion: "There is disagreement on the value of the space flight experience. . . . The task force believes the only way to decide is to take a small step which has value in itself and see what light it sheds on what the next step, if any, should be."

So NASA pressed on, forming a Space Flight Participant Program, which would allow at least two private citizens — the George Plimptons of space travel — to fly on the shuttle in 1986, more later. They would train for four months, fly for a few days and then crisscross the country to share the wonders of their unearthly vacations. The first pioneer, NASA had decided, should rise from the ranks of education. Beggs would recommend to the president that a teacher should reach for the stars.

Teachers, after all, would not spill paint in the orbiter, as an artist might. They would not write poetry that soared over the heads of middle Americans, nor would they beam back the blow-dried image of a television anchorman. They were everyday people, symbols of tradition and innocence, underpaid, and compassionate leaders of the nation's children. Shooting a teacher into space, observed Rush-

110

worth Kidder, a reporter for the *Christian Science Monitor*, was "a conscious decision to put a dove among the eagles."

Reagan announced the plan on his first full day of work after winning renomination as the Republican candidate for president.

"When that shuttle lifts off, all of America will be reminded of the crucial role teachers and education play in the life of our nation," he said at Jefferson Junior High School, eight blocks from the White House. "I can't think of a better lesson for our children and our country."

Others could. Denouncing Reagan as "America's No. 1 Scrooge on education," the National Education Association had recently endorsed his opponent and NASA critic, Walter Mondale. Now it condemned the president's teacher-in-space announcement as "a gimmick," an attempt to camouflage problems in the schools and pacify NEA members, who accounted for about 80 percent of the nation's two million teachers.

"We don't need to send a teacher into space," said Mary Hatwood Futrell, the union's president. "We need to send teachers into well-equipped classrooms."

Beggs denied there had been political pressure to choose a teacher. NASA had given the president a list of recommendations, Beggs explained, and the teaching profession had topped the list. But the critics complained that education should have topped more of Reagan's lists.

"Teachers have been scandalously neglected in this country," said Dr. Ernest Boyer, president of the Carnegie Foundation for the Advancement of Teaching. "Putting one into orbit is a powerful, symbolic message. Let it be the beginning of America's commitment to them, not the end."

Less solemnly, Russell Baker, a columnist for the *New York Times,* joined the chorus of skeptics.

"It's hard to imagine a good teacher who would be anything but a useless curiosity among a working space crew," he wrote. "Worse, the media malarkey in which this master teacher would be draped would turn a good schoolteacher into another useless celebrity adept at breakfast-time chatter with David Hartman and Jane Pauley, but too world-weary to pound the desk effectively at a slovenly student's misreading of Caesar. . . . This would be a sad end for a good teacher."

Christa's hometown paper, the *Concord Monitor,* ran the news — REAGAN WANTS TEACHER IN SPACE — on page one next to a picture of Judith Resnik, the wind catching her long black hair as she climbed out of a supersonic training jet at the Kennedy Space Center two days before the twelfth shuttle mission, her first. Picking up the paper on her front porch, Christa glanced at the story, then at the photo. She had included Sally Ride in her women's history course when Ride became the first American woman in space a year earlier. Now a second woman was on her way, and soon — it was hard to believe — a teacher would follow. There may be hope yet, she thought.

The entry rules for NASA's contest seemed so simple, so downright accommodating, that she imagined the original astronauts wincing with envy. Candidates needed only to be American citizens, have five years of classroom experience and suffer no illness that would threaten their safety or the safety of the crew. Their vision needed to be correctable to 20/40 in their better eye (Christa's contact lenses were permissible), they needed to be able to hear a whisper at three feet (hearing aids were also permissible), and their blood pressure could be no higher than 160/100. That was it. Not even an age requirement.

"We're not looking for the teacher who's the strongest or the tallest or the most flexible," said Ed Campion, a NASA public affairs officer assigned to the teacher-in-space race. "We're not looking for Superman. We're looking for the person who can do the best job of describing their experience on the shuttle to the most people on Earth."

Finding Superman almost seemed easier. With more than two million potential teachernauts, all of them trained communicators, NASA first needed to clip the wings of anyone who had only a passing interest in space flight. A twenty-five page application helped. Fat with instructions, essay questions, requests for background information and recommendations, the application required as many as 150 hours for a serious contender to complete. And there was an application just to *obtain* an application.

"It was a self-selection process," said Alan Ladwig, the director of the Space Flight Participant Program. "The bottom line was: how badly do you want to fly?"

Christa threw herself into the new school year and for nearly three months forgot about the contest. Then, on the weekend before Thanksgiving 1984, she attended the annual meeting of the National Council of Social Studies at the Washington Hilton and noticed a display table covered with request forms for the teacher-in-space applications. She glanced about the room and when no one was looking, she tucked a thick stack into a folder. Christa returned to Concord High, scattered all but one of them among friends and mailed hers to NASA. A week later, her application, a glitzy blue-and-silver packet with a picture of the shuttle on the cover, arrived in the mail. The deadline to apply by was February 1, 1985. She shoved it in a drawer.

"No sense kidding myself," she said. "If there was anything I was going to put off to the last minute, that was it. It looked ten times worse than a term paper."

So the race started without her. Even before NASA accepted its first application on December 1, starry-eyed teachers had tried scoring points. More than forty of them had joined Ladwig in late October for a weekend at the U.S. Space Camp in Huntsville, Alabama, a fifty-acre high-tech amusement park where they had pulled on astronaut blues, flown in a mock shuttle and sampled the sensation of weightlessness.

"With what I've experienced this weekend, I feel like my dream just might come true," said Michael DiSpezio, a teacher from Cape Cod. But DiSpezio nor anyone else in the group survived the first cut of NASA's talent search.

Other teachers tried a more scholarly approach, combing the space sections of their libraries, poring through aeronautic journals, renting videocassettes of *The Right Stuff*, writing to NASA for pamphlets and pictures, hoping their homework would help them impress the judges. A few, like Chris Owen of Nevada County, California, tried to dazzle the space agency with his flair for public relations. Owen, a thirty-two-year-old woodshop teacher, spent the first week of 1985 conducting a grass-roots campaign for himself on the streets of New York. He strolled hundreds of city blocks and stood on dozens of sooty subway platforms, passing out buttons that said "Owen in Orbit — First Teacher in Space." Owen lost in the preliminary round as well.

At Riverhead High School on Long Island, students formed the Bob in Outer Space Club to campaign for their science teacher, Bob Jester. They fanned out, some to collect signatures for a petition, others to organize bake sales, car washes and candy sales to raise enough money to travel to the state capital and present the petition to the legislature. They wore "Bob in Outer Space" buttons when they arrived, but Bob was also eliminated in the first round.

In Washington, thousands of pieces of mail

arrived from children interested in sending their teachers into space. A seventh grader in Fresno, California, wrote that her English teacher was "not afraid of heights, as I am, and gets along well with others," even "seventh graders, who are not the easiest people to get along with." Several students said their teachers would show Martians how to do the moon walk, and the entire student body of a Denver elementary school signed and delivered a forty-foot scroll to tout their candidate. Many children simply sent a few kind words in support of their favorite hall monitor. Others were less cordial.

A boy from Sheridan, Wyoming, complained to the president that his teacher was "mean. He makes us take tests too often. He doesn't let us talk in class. Please take him. I want him out of here." Another boy wrote, "Send my teacher to Mars. That's where her relatives are."

No one wrote about Christa, and Christa spent little time thinking about the contest. She might have forgotten about it altogether had not Steve started prodding her soon after Christmas. Time is running short, he reminded her. This is a don't miss. Don't let it slip away. Go for it.

Steve was a gifted writer and Christa was

not, so she cringed at the thought of answering eight essay questions, among them "Why do you want to be the first U.S. private citizen in space?" and "Describe your teaching philosophy." She stalled, then stalled some more, until finally, ten days before the deadline, she could stall no longer. She pulled out the application.

Those were the quiet nights of winter, the icy, New Hampshire nights when the storm windows were locked tight and the snow on the street muffled the sound of passing cars. For a week Christa went home from school, cooked dinner, read Caroline to sleep and tucked Scott into bed with Rizzo. She corrected papers, said good night to Steve and brewed a cup of tea. Then she wrote. And wrote.

"I'm so sick of looking at these things," she told Eileen O'Hara on the seventh day. "I've rewritten each of them four times."

Christa wore a happier face in class. In her breezy style, she explained that writing was sometimes a painful exercise, but that history had taught her people sometimes have to suffer to grow. She told her students that applying for the space flight was a chance to test her potential, to look beyond herself, to dream. Everyone needed to dream, she said, and she reminded them of what her father had told her as a child: "Some people never fail, but some never try."

Some of them rolled their eyes; others learned.

"I used to think only teachers in first grade and second grade affected your outlook on life," said Audra Beauvais, the superintendent's daughter and Christa's student in "The American Woman." "Then I took Mrs. McAuliffe's course, and she changed all that. It was like she discovered something new every day, and she was so excited about it that it got the rest of us excited, too. She made us really want to live life."

Christa explained in one of her occasional progress reports to her students that answering most of the application was no harder than writing a hall pass. She had supplied background information and lined up recommendations from a colleague (O'Hara), an administrator (Charles Foley) and a community member (Arthur Robbins, a Concord district court judge). She had obtained permission for a year's sabbatical (if she won, NASA would pay her salary of $26,681, plus living, travel and child-care expenses), and she had signed a contract in which she surrendered her rights to many government benefits, including life insurance. Nothing nerve-racking, she said, nothing as gnawing as the essay questions, particularly the one that asked her about the project she would conduct on the six-day mission.

She had thought of an idea — to keep a diary — and had started writing an essay about it, but she worried privately that the judges might consider the proposal too frivolous. She hesitated telling too many people — seven other Concord teachers were competing — but she needed to talk to someone, and one afternoon she cornered a colleague, Ron Brown, as he drank coffee in the social studies office.

"What about a journal?" she asked near the end of a rambling soliloquy. "It wouldn't take up any space, and I wouldn't get in anyone's way. I've kept one for years, you know, and it would help me remember everything so I could share it when I got back. Don't you think it would work? Sure, that's what I'm gonna do. That's my project. A journal."

An ordinary person's letter from space.

Brown looked up to nod his approval, but Christa was rushing out the door, her hands full, headed who knew where. Her deadline was three days away.

"I don't know if I'll make it," she said in

class the next morning, "but I'm trying and I'm still having fun."

She revised the essays a fifth time and asked O'Hara for help, partly because O'Hara could type the final draft at 105 words a minute, also because she needed an editor. Huddled over a word processor in Steve's law office, they spent several hours polishing Christa's handwritten prose. Snow fell outside, but they hardly noticed it as they labored over a misplaced preposition here, an undeveloped idea there. Once, Steve sneaked up behind them and pretended to read over Christa's shoulder. Suddenly, he laughed.

"You're going to say *that?*" he said in mock disbelief, backpedaling before Christa could catch him.

A night later, Steve sat beside her at home and read her application, first the essay on why she wanted to be the first private citizen in space.

"I remember the excitement in my home when the first satellites were launched," Christa wrote. "My parents were amazed and I was caught up with their wonder. In school, my classes would gather around the TV and try to follow the rocket as it seemed to jump all over the screen. I remember when Alan Shepard made his historic flight — not even an orbit — and I was thrilled. John Kennedy inspired me with his words about placing a man on the moon, and I still remember a cloudy, rainy night driving through Pennsylvania and hearing the news that the astronauts had landed safely.

"As a woman, I have been envious of those men who could participate in the space program and who were encouraged to excel in the areas of math and science. I felt that women had indeed been left outside of one of the most exciting careers available. When Sally Ride and the other women began to train as astronauts, I could look among my students and see ahead of them an ever-increasing list of opportunities.

"I cannot join the space program and restart my life as an astronaut, but this opportunity to connect my abilities as an educator with my interests in history and space is a unique opportunity to fulfill my early fantasies. I watched the Space Age being born and I would like to participate."

No poetry, no hyperbole, Steve thought, just a simple request for a two-million-mile ride of a lifetime. Nice touch, he told himself. Then he read her proposal for a shuttle project.

"In developing my course, <u>The American</u>

Woman, I have discovered that much information about the social history of the United States has been found in diaries, travel accounts and personal letters. This social history of the common people, joined with our military, political and economic history, gives my students an awareness of what the whole society was doing at a particular time in history. They get the complete story. Just as the pioneer travelers of the Conestoga wagon days kept personal journals, I, as a pioneer space traveler, would do the same.

"My journal would be a trilogy. I would like to begin it at the point of selection through the training for the program. The second part would cover the actual flight. Part three would cover my thoughts and reactions after my return.

"My perceptions as a non-astronaut would help complete and humanize the technology of the Space Age. Future historians would use my eyewitness accounts to help in their studies of the impact of the Space Age on the general population."

Steve, like his brother, Wayne, a Navy instructor pilot, had always dreamed of flying. From the time he was old enough to build model airplanes, Steve had wanted nothing more from life, but his poor eyesight had grounded him. Doctors told him they could not correct his 20/400 vision. Specialists told him they knew of no operation that would help him, and he made his last attempt to earn his wings when the military relaxed the vision requirements for helicopter pilots at the height of the Vietnam War. Even then his eyes were too weak.

Steve had once arranged for Christa's mother to take a flying lesson in the skies above Concord as a birthday present. Now Christa was working toward what he described as "the ultimate flying experience," and as he read the rest of her essays — describe your communications skills, describe your community involvement, describe your professional development activities — he realized she was on the edge of his dream. When he finished, he laid the application on the coffee table and said nothing for a while. Then he smiled at her.

"Where is this woman?" he asked. "I want to marry her."

The next morning, a steel-gray, subfreezing Friday, Christa rushed to the post office before school and mailed her application. It was February 1, the last day to apply.

Two weeks later, she and Steve took the children to Disney World, where Christa posed for an artist who drew a caricature of her aboard the space shuttle. The caption said, "The adventures of Christa — The teacher in space!!"

Meanwhile, the selection process had started. More adept at selecting astronauts than private shuttle passengers, NASA intended to steer clear of the process until the field of candidates had been trimmed to ten. Clerks were to review the applications, eliminate anyone who failed to meet the eligibility requirements and forward the rest to the Council of Chief State School Officers, which had formed selection committees for each state as well as the Bureau of Indian Affairs and U.S. schools overseas. Each committee was to nominate two candidates who fit the profile of the perfect teachernaut: a gifted communicator, a strong role model, an angel from Madison Avenue.

"It was nothing like the Teacher of the Year competition," said Dr. William Ewert, chairman of the New Hampshire selection committee. "NASA wasn't looking for the teacher who communicated the most effectively with students. It wanted the teacher who communicated the most effectively with *everyone*."

More than 45,000 teachers requested applications, but fewer than 11,500 applied, including about 1,000 in California, nearly

800 in New York, only 79 in New Hampshire.

"Look no further," many of them had written. "I'm the one!"

The applicants included authors, Rhodes scholars, doctors, a former pro football player, pilots, hot-air balloonists, mountain climbers, skydivers, marathoners, a champion sailor, a ventriloquist, a man who built a liquid-fuel rocket in high school, a woman whom NASA had considered for the astronaut corps. None of them was from New Hampshire.

"In New Hampshire, we had a lot of teachers trump up shuttle projects that turned out to be simply bad science experiments," Ewert said. "We really had no trouble finding seven [state] finalists."

On April 16, 1985, he summoned the seven for interviews at the Walker Elementary School, an aging red-brick building with a chipped blacktop playground a mile from Christa's house and a block from her church.

"I hope we get this over with soon," she told Susanne O'Brien, another finalist from Concord, as they waited outside the principal's office. "One way or the other, I can't wait to find out."

"Neither can I," O'Brien said.

Christa went first, and O'Brien was out of luck. One of the judges, Rosemary Duggan, was so impressed by Christa that she settled on her after a single question: "What in your career has caused you to grow or change professionally?"

Pausing for a moment, Christa looked about the room and breathed deeply. Then she began. In her second year of teaching, she said, 80 percent of her students had been black, poor and culturally isolated. They had lived just outside Washington, a nerve center for international relations, but they had known nothing about the world beyond their backyards. Vowing to change that, she had contacted the Peace Corps and committed them to a fund-raising drive for a new school in Africa. They had raffled a television, held bake sales, a car wash and a talent show, and had raised enough money to build a one-room cinder-block school in a rural region of Liberia where there had been no school. They had learned about geography, international relations, community activity and, four years before Alex Haley's book, they had learned about roots. Christa had learned what a difference the extra effort could make in education, and she had saved a picture of the school as a reminder.

"She could have just read the chapters and taken the easy way out," Duggan said. "But she was someone who knew why she was in the teaching profession and why she loved it. She could communicate that, and better yet, she was someone who could hold up on *Good Morning, America*."

Duggan's colleagues agreed.

"Some of the teachers had better résumés, but I think we knew right away Christa was a candidate who could make the final ten, even number one," Ewert said. "She had that girl-next-door quality, a kind of wholesome American look."

The phone rang as Christa cooked supper that night. It was Ewert to tell her she was one of the state's two nominees, a unanimous choice.

"That's great!" she said, suddenly a soprano. "Can you hold on? I have to tell my husband."

The next morning she arrived late, as always, to her first-period women's history class, but not as late as usual. A knot of students met her at the door.

"How'd you do?" one of them asked.

"I made it," she said, grinning, her hair still wet from her morning shower.

They applauded, and although the rest of her students had not heard the news, they joined in. For a moment, they believed they were clapping because Christa had arrived so soon after the bell had rung. Then they understood.

"Hey, atta way, Mrs. McAuliffe!" they shouted. "Don't stop now!"

"Can you believe it?" she said. "I still haven't come down yet."

The other New Hampshire nominee was Robert Veilleux, an astronomy and biology teacher from Manchester. Within two weeks, he and Christa were to undergo medical tests by the Federal Aviation Administration and to answer three questions in a videotaped interview that would be used by judges in Washington in the next round of the selection process. The physicals were no problem. The interviews, at least for Christa, were.

On the morning she faced the state selection committee, Christa had conducted a television interview for WNHT, a small Concord station, her first television appearance since the age of four. She had been stiff, she thought, uneasy. She could perform for a roomful of adolescents, but this video madness . . . this was something else.

"It was strange, really, because she spoke in perfect, twenty-second bites," said the reporter, Martha Cusick. "She was a natural, but I'm not sure she knew it at the time."

She had no idea. Convinced the videotapes could be decisive and that she knew nothing about the subtleties of television — the use of body language, how much makeup to apply, the correct colors to wear, the proper pacing and tone of her voice — Christa confided to her jogging partners that she was in trouble. She was so worried, she said, that she had talked to herself on the way home from the tennis court, attracting strange looks from motorists who had stopped next to her at the traffic lights.

One of the joggers, Barbara Jobin, offered to help. Christa could rehearse in front of her husband's video camera, she said. She could conduct her own private screen test.

With Scott and Caroline in tow, Christa climbed into her Volkswagen bus one night after dinner and drove down the Hill to Jobin's apartment, a fashionable second-floor walk-up with high ceilings and teak furniture three blocks from the state house. Scott arrived in his red pajamas, Caroline in her Care Bears nightgown, Christa with her arms full of clothes, her purse stocked with newly purchased cosmetics. Jobin's husband, Roger, stood ready with his video camera.

He waited, however, while his wife coached Christa in the art of cosmetics. When Christa emerged from the lengthy makeup session, her lips were glossed pink, her eyes were shadowed and her cheeks powdered. She wiggled into the interview seat, a wooden chair they had placed in front of a fabric wall hanging, and smiled when Roger clicked on his camera.

"Hi, my name is Christa McAuliffe. I teach at Concord High School and I would like to describe my shuttle project," she said. "The project I've chosen is to keep a journal . . . "

She wore a black sweater vest and a light blue shirt with a white collar. She spoke easily, her eyes wide, her head bobbing with enthusiasm. She ended in a flourish, explaining with charm and confidence that she "would like to bring back my perceptions of what an ordinary person sees on board the shuttle." Then, smiling, she threw out her hands and cried, "Cut!"

She watched the tape immediately on a television in the living room. The lights were too low, she thought, but that was no problem. She worried more about her delivery — her eyes had wandered — and her wardrobe — the white collar would have to go. She retreated to the bathroom and returned a few minutes later wearing a simple, dark blue dress.

"Here we go," she said. "Take two! Hi, my name is . . ."

Just then Caroline darted in front of the camera. Eager to imitate her mother but not quite sure how, Caroline had considered singing a song. Now she decided it would be more fun if her mother sang it.

"Mommy, sing Happy Birthday," she said.

The camera whirred and Christa looked solemnly into the lens.

"Hi," she said, "my name is Christa McAuliffe and" — smiling suddenly — "Happy birthday to you, happy birthday to you, happy birthday dear . . ."

Caroline squealed, dashed toward the bathroom and announced she was going to put on makeup. Scott curled up on the couch with his *Super Sports Trivia Book* and continued reading, trying desperately to ignore the proceedings. Christa turned her attention back to the camera.

"Hi, my name is Christa McAuliffe . . ."

After posing in each new change of clothes, Christa studied the videotape and fretted. Was the blue dress too dull? Would the white blouse blend in with a white background? Which of her four vests was the best? Would the camera focus only on her head and shoulders? Where would her interviewer sit?

Should she look at the interviewer? Or at the camera?

By the fifth screen test, it was nearly 11:00 P.M. and Scott and Caroline were asleep on the couch. Roger Jobin's hands were getting tired, and Christa was getting giddy.

"Hi," she said, "my name is Ronald Reagan . . ."

When Ewert picked her up a couple of days later for the forty-five-minute ride to a television studio at the University of New Hampshire, Christa wore a gray sweater vest, a light blue shirt with a button-down collar and plenty of makeup. She mentioned apologetically that she rarely wore so much makeup. Then she pulled some notes from her pocket and said she wanted to study. She rode the rest of the way in silence, her lips moving as she rehearsed her lines.

The interview was to consist of three questions — two she could prepare for and one that required an unrehearsed answer. None of the answers could exceed ninety seconds, and Ewert, the interviewer, flipped a coin to determine whether Christa or Veilleux would face the klieg lights first. It was Christa.

"Please describe the project you outlined in your application for your time in the space shuttle," he asked her.

Instead of looking at Ewert or the camera, she glanced about the studio as she answered each question, imagining herself trying to hold the attention of a last-period class on the first day of spring. The camera focused only on her head and shoulders, and she made the most of it, filling the frame with enthusiasm, talking as loudly with her eyes as she did with her words.

"Having been a history teacher for quite a few years, I've been very aware of the fact that social history — the history for the majority — is often unknown or forgotten," she said. "Because it's very easy to chronicle military, political and economic history, and social history is a little more difficult, the common man is often pushed aside."

Each semester, she explained, she sent her students into the community to ask World War II veterans what it had been like to fight in the Philippines; Vietnam veterans what it had been like in their war; senior citizens about the Depression; common people about their lives in the twentieth century. The students shared their findings in class, she said, just as she intended to share her impressions of space travel.

"The shuttle flight is certainly going to be a dream and a thrilling experience for me," she said, "but it would be even more wonderful to share it with the people back home."

Unlike many of the other candidates, Christa brought no props, quoted no Shakespearean verse, wore no American flag on her vest. Her only symbol was herself, an ordinary teacher eager to share an extraordinary experience with ordinary people. She had yet to falter.

Ewert's second question: "This is a major time commitment. If you are selected, what impact do you see this having on your life beyond the classroom?"

Christa breathed deeply and smiled.

"Well, certainly it will change my life," she said. "I have been a classroom teacher for these years, but I've also been an adventurer and I've always wanted to try something new. As a historian, to have the opportunity to embark on such an adventure, I certainly wouldn't be able to say no. It *will* change my life.

"I think it would be great to travel and share the experience and see people, but I also think it would give me an opportunity to go back into teaching with renewed enthusiasm, having really seen history in the making, which is so important when you're trying to bring these kinds of things across to the students.

"My family is excited about it, and the students at school are certainly excited about it. I think it will give me an opportunity to meet a lot of people and to share the space age, which right now is kind of removed. We only see little clips of it in the newspaper, and we really don't understand the whole ramifications of what's going to be happening in the next twenty years."

So far so good. She had connected on the first two questions, but here came the curveball, the question she feared the most, the one that left candidates stuttering from Anchorage to Atlanta. Ewert paused. Then slowly he said, "Describe your philosophy of living."

"My . . . philosophy . . . of . . . living," Christa repeated, even slower, as if to say, Steve, you'll never believe this one: My . . . philosophy . . . of . . . living.

She had ninety seconds to explain what some people do not understand in a lifetime. She swallowed hard.

"Well, I think my philosophy of living is to get as much out of life as possible," she said. "I've always been the type of person who is very flexible and has tried new things. I feel that you need a little bit of organization, but I also think it's important to connect with people.

"I think the reason I went into teaching was because I wanted to make an impact on other people and to have that impact on myself. I think I learn sometimes as much from my students as they learn from me. Being in an educational field has also been the type of thing where I can go out and see people later on in the community, see them in their twenties and having their families and growing up, and that has given me a good feeling.

"My philosophy of living, I suppose, is to enjoy life and certainly to involve other people in that enjoyment, but also to be a participant and to enjoy all the things we have in this country."

She had hardly unhooked her microphone when she asked to replay the interview. Watching intently, Christa worried about her makeup and winced at her reaction to the final question, but she detected no serious flaw and that night told Steve she had been satisfied with her performance. Still, she was surprised when NASA officials told her in July that the interview was one of the overriding reasons why she was chosen for a ride in space. She thanked Roger and Barbara Jobin with a jar of homemade jam.

The weeks after the interview passed quickly, and soon it was the last day of classes at Concord High. Christa told her students she was as nervous about entering the next leg of the space race as they were about facing their final exams the next few days. She told them to go for it, push themselves, and that she would do the same. She wished the seniors well and said she looked forward to seeing everyone else in September. And if they signed a sheet of paper as they left their exam rooms, she said, she would send them postcards from Washington, Houston, wherever the space adventure took her. They applauded and wished her luck. Some of them gave her a parachute as a going-away gift. Others shook her hand. They all said good-bye. It was Christa's last day as a classroom teacher.

CHAPTER FIVE

Here they were in Washington, all the way from Honolulu and Hattiesburg, Tucson and Texarkana, an island in the South Pacific, an Eskimo village in the Arctic Circle, 113 teachers competing for an all-expenses-paid trip around the world 97 times. And here was James Beggs, the man who would choose the flier, quoting the Phyrgian philosopher Epictetus to them as they digested their filet mignon at NASA's welcoming banquet in the grand ballroom of the L'Enfant Plaza.

"Make the best use of what is in your power," Beggs told the teachers. "Take the rest as it happens."

Like Christa, most of them intended to do just that, to make the best of the six-day talent contest: to showcase themselves as they attended workshops, press conferences and movie premieres; mingled with astronauts, congressmen and the president; sipped cocktails with a television star; rode a party boat down the Potomac. They would face the judges on the fifth day, and on the seventh — June 28, 1985 — the 113 would become 10. Three weeks later, there would be 1.

"And should that lucky one" — unable to take the rest as it happened — "get cold feet at the last minute," said Beggs, his voice thundering through the loudspeakers, "he or she need only give me a call and I'd be happy to fill in!"

The launch of mission 51-L was scheduled for January 22, 1986, exactly seven months away. (NASA used the 51-L designation to signify by the "5" that the mission was budgeted in the 1985 fiscal year, by the "1" that it would lift off from the Kennedy Space Center and by the "L," the twelfth letter of the alphabet, that it would be the twelfth mission of the fiscal year.) If NASA had its way on that January morning, the world would watch a teacher fly away from the Florida coast on one of its billion-dollar space trucks — the shuttle *Challenger*. Earlier, *Challenger* had ferried the first American woman and the first black into space. It had recorded the first nighttime launch, the first night landing, and it had served as the springboard for the first untethered space walk. The first celestial marketing battle had even occurred on its middeck, a battle Coca-Cola won when the crew, testing two experimental space cans, decided Coke was it and opened Coca-Cola's can before Pepsi's.

Now 113 teachers wished they were *Challenger*'s next first. And here was the pilot, Michael Smith, following Beggs to the podium to tell them what it might be like. Not that he really knew.

Smith, a chicken farmer's son, had grown up in Beaufort, North Carolina, not far from where the Wright Brothers first flew. Like Christa, he had read the biographies of the original astronauts and learned they had been test pilots and military pilots, some of them combat pilots. Determined to become one of them, he had pursued the same course. He had made his first solo flight on his sixteenth birthday, attended the U.S. Naval Academy, flown more than two hundred missions as a fighter pilot in Vietnam and completed the Navy test pilot school. When NASA had called him in 1980, the waiting period for astronauts to fly was five years. His wait was nearly over. Smith was forty years old.

Tall and lean, Smith had the look of the early astronauts but the manner of a humble farmer: a gentle wit, a strong sense of purpose, an ego that needed no stroking. He had been highly decorated in Vietnam — the Navy Distinguished Flying Cross, the Vietnamese Cross of Gallantry with Silver Star, thirteen Strike Flight Air Medals and more — but his own family had not known about them until he spoke ten years later at a college dedication and someone read his Navy biography as an introduction. Rather than the "right stuff," his friends said, he had been blessed with the "nice stuff."

Smith told the teachers in a soft country drawl that he could hardly wait "to get on orbit and get the secret handshake." He described the cargo his crew would deploy in space — a free-floating laboratory to study Halley's comet and a communications satellite that would boost NASA's contact with orbiting shuttle crews from eleven hours a day to more than twenty a day. He talked about his crew mates — Commander Francis ("Dick") Scobee, and Mission Specialists Ronald McNair, Ellison Onizuka and Judith Resnik (Gregory Jarvis had not yet been assigned) — and explained that each of them had flown before. If space flight was half of what they claimed it was, Smith said, he and a teacher were destined for a spectacular adventure, a journey that Sally Ride had described as "an E ticket [good for every attraction] at Disneyland."

Christa sipped a cup of tea and scribbled notes.

"All I really knew about space travel before that was what I had seen on *Star Trek*," she said. "Suddenly, hearing about it from an astronaut, it started to come alive."

Christa, late as usual, had arrived at the L'Enfant Plaza that afternoon with a reservation but no room. Teachers had descended on the hotel by the dozen: two from each state as well as from American schools in Kuwait and Jakarta, Puerto Rico and the Virgin Islands, Germany and Guam. Two were to represent the Bureau of Indian Affairs, but one of them had left teaching and been disqualified. Still, for one night there were too many contestants and too few rooms. NASA found accommodations for Christa and seven others at the Hyatt Regency across town.

As she straightened out her clothes that night, including eight new outfits her parents had bought her, Christa was tormented by her first impressions of the competition. The other teachers had been more striking than she had ever imagined. She had arrived in Washington hoping to make the final 10, but now she believed her chances were better of making, say, the top 110. She slumped to the edge of her bed and started to read their capsule biographies. Soon she could read no more. She called home, her voice flat with disappointment.

"Steve, these people are doctors and authors and Fulbright scholars and teachers of the year and a woman who climbed the Himalayas and . . . I'm out of my league. I haven't got a chance."

"Hang in there," he said. "You're doing fine. You wouldn't be there if you didn't have merit. Just relax and have a good time."

She fell asleep with the biographies on her bed, comforted, at least for the moment. The next morning Christa returned to the L'Enfant Plaza for a jolt of reality.

Like the rest of the 113, Christa knew a ride in space would cost more than a few days in the classroom. She knew the winner would fly for six days and spend a year afterward stumping for the space agency, but she was confused about exactly what NASA expected of the winner *before* the flight. The application had been vague. A Project Timeline had said: "Fall 1985 — 120 hours of training." The Selection Criteria had mentioned "preflight, flight and post-flight activities (a commitment of approximately 18 months)." And the Training Requirements had provided a bit more detail: "The primary and back-up

144

teachers will undergo approximately 120 hours of training during the eight weeks prior to flight, as well as certain testing requirements at an earlier stage." Nothing had indicated the winner would spend much more than eight weeks away from home before the flight.

Now here was Alan Ladwig, manager of the Space Flight Participant Program, announcing that the winner would report for training in early September, more than four months before lift-off.

"It's a big commitment," he said, "and we don't want to get anybody divorced over this."

Did he say divorced?

"Everybody, please understand what you're getting yourselves involved with, and don't say we didn't warn you. We can't afford to have you come up in the middle of training and say, 'Well, I've had second thoughts about this and I want to go back to my hometown.' That's not part of the deal."

No longer would the winner leave home around Thanksgiving. Now it was the first week of school in September. What else did NASA have in mind? Would the winner get home for Christmas, for heaven's sake? The stakes had suddenly gone higher, and Christa, like most of the 113, began to consider more seriously the long separation from her family.

Did he say *divorced?*

The teachers split into four groups after Ladwig's lecture, each group named for a shuttle: *Atlantis, Challenger, Columbia* and *Discovery*. They spent the afternoon in workshops — "Living Aboard the Space Shuttle," "Flying Aboard the Space Shuttle," "Looking Toward the Heavens" — and passed the evening dining and dancing on a cruise down the Potomac. Although Christa belonged to the *Challenger* group, she spent most of her free time with a teacher from the *Columbia* team, Charles Sposato, an English teacher at Farley Middle School in her hometown of Framingham. They each had attended Framingham State, and they had several other things in common.

Sposato's wife, Mary, knew Christa and her husband from Marian High School, where she had graduated a year before them. Christa's mother knew Sposato from her work as a substitute art teacher at Farley, and Christa's brother, Steve, a lawyer in California, knew Sposato from the years Steve had run the audiovisual department at Farley to pay his way through night law school. Her brother knew Sposato so well, in fact, that when he heard Christa was in the same contest with him, he told her, "Forget it, you don't stand a chance against that guy."

Christa and Sposato talked as their colleagues danced, the city lights shimmering on the river, a summer breeze rolling across the main deck. They talked about their families — Sposato had three children — about Washington, Christa's years there as a teacher and Sposato's visits as a seminary student. They talked about Christa's brother (they called him Steve 1) and her husband (Steve 2). They talked about teaching and their shuttle projects — Sposato had also proposed to keep a journal. And when they talked about jogging, they agreed to meet the next morning at dawn.

An idling taxi and a flock of chirping birds greeted them at 5:30 A.M. They stretched against the hotel and set out, at Christa's suggestion, for the Jefferson Monument, running three miles through the morning haze, dodging an occasional car, chatting, unaware that on the other side of the planet, the shuttle *Discovery*, with three Americans, a French test pilot and Saudi prince on board, was about to begin its fiery descent through a wall of gravity on the last leg of a successful flight, the eighteenth by a space shuttle.

Back in her hotel suite Christa ate a room-service breakfast and watched the news. An update on *Discovery*'s return followed a series of reports from Lebanon, where forty Americans had been taken hostage on a commercial airliner. Even then, the news of *Discovery*'s dangerous reentry into the Earth's atmosphere was dwarfed by a press conference aboard the shuttle the day before. A reporter had asked the Saudi prince what the Earth looked like from the heavens.

"Looking at it from here, with trouble all over the world, not just the Middle East, it looks very strange as you see the boundaries disappearing," the prince had said. "Lots of people who are causing some of these problems ought to come up here and take a look."

Christa smiled and nodded her approval. An hour later she applauded when a NASA official announced that *Discovery* had touched down on a dry lakebed at Edwards Air Force Base in the Mojave Desert. The ninety-eighth, ninety-ninth and one hundredth American to enter space had returned home safely.

Christa and Sposato continued their morning runs, visiting the Lincoln Memorial, the Vietnam Veterans Memorial, a different historical site each day. They became close friends and confidants as the tension of the week mounted, but on this first morning they were more concerned with arriving on time for their first class, a lesson in dealing with the price of celebrity.

For a year, maybe more, the winning teacher would be the most visible spokesman for a space agency that relied on public support for its $7.6 billion budget.

"You will be considered the expert, so you will have plenty of opportunities to speak out, probably more than you'll want," said Frank Johnson, NASA's public affairs director. "You also will be giving more autographs than you want and making a hell of a lot more appearances than you want. That is a fact of life."

The unstated fact, of course, was that the life of the first private citizen in space would no longer be private. And that meant they better be prepared for the press.

"Local press, national press, print media, TV types, radio types — all of them are going to come at you from many different directions," said Johnson, who claimed to have the best public relations job in America. "Nobody's going to be out to destroy the space program, but, you know, when you get fingers into the pie, sometimes things can get messed up."

To help the teachers avoid such a mess, Johnson had brought along Walter Pfister, a broadcast journalist who had coached most of the shuttle astronauts and NASA's top management.

"We're not going to make Barbara Walters or Walter Cronkite out of you," Pfister said.

He simply wanted to ask a few questions and offer a bit of advice.

"How many of you have dealt with the press in the last two months?"

They all raised their hands.

"And how many of you have had negative experiences?"

Three, maybe four, not Christa.

He told them how to eliminate those experiences. Relax, he said; speak clearly and concisely, avoid technical jargon, cooperate with your interviewer and, most of all, know your audience. On a morning talk show, for instance, understand to whom you are speaking: a middle-aged accountant in Portland, Oregon, who never watches television but is home with the flu; a truck driver who just made a run from Chicago to Orlando and is winding down with a beer and a few minutes of television; an eighty-three-year-old widow in Queens, New York, who watches television all the time because she has nothing better to do. Remember you are an ordinary person speaking to ordinary people.

"Be conversational," he said. "Be yourself. It works!"

Christa knew that, of course, and she was prepared for a press conference later in the morning. But first came a briefing on the shuttle's safety record and its most persistent problem — scheduling delays.

NASA's inability to stay on schedule made it increasingly difficult to sell cargo space on the shuttle, said Chester Lee, the director of NASA's customer service division. By flying less than a third of the forty-eight missions a year it had projected in 1972, NASA had been forced to raise its payload price from a projected $270 a pound to more than $5,200 a pound, inviting stiff competition from European and Japanese space agencies. Lee wanted to make one thing clear, however.

"We want to maintain our schedule, but that comes after safety," he said. Before every launch "we have numerous reviews, including flight maintenance reviews, where top management sits in to be sure that everything possible has been done" to prevent unacceptable risks on the "most heavily instrumented vehicle in the world."

He said the shuttle was equipped with

"redundancies," space talk for backup systems, and explained that it was impossible to build redundancies for some parts of the craft. NASA ensured the safety of those parts, he said, by building them to resist twice the amount of stress of a normal space flight.

"So you don't get a burn-through in the solid rocket boosters," for example, "which would not be good, they have about twice the amount of insulation. They have the strength and capability to withstand about anything you can conceive."

The extra protection was necessary, he said, "because we're very concerned about the first two minutes you're on the solid rockets. If one of those rockets goes, why, it's pretty bad."

No one asked how bad.

The teachers paraded in single file from Lee's lecture into the press conference, where, like students at high school graduation rehearsal, they lined up on risers, twenty-three to a row, shoulder to shoulder, and posed for the cameras. Most of them taught in small towns, seven in elementary schools, twenty in junior highs, seventy-seven in high schools. They had taught an average of fourteen years and half of them taught science. There were fifty-nine men and fifty-four women who ranged in age from twenty-seven to sixty-five.

Television cameras whirred as three dozen reporters asked questions — "What's it like?" "What are your chances?" "What do you think NASA's looking for?" — and the teachers, like eager students, raised their hands. The session lasted a half hour, but Christa, the twenty-second person in the third row, was lost in the sea of hands and was never called upon. She grew anxious as others, like Sophia Clifford, a chemistry teacher from Birmingham, Alabama, scored points with the press.

Clifford, standing in the front row, said she believed she might win the shuttle ride because a fortune cookie she had cracked open at a Chinese restaurant in Birmingham had said, "You've been promised a ride on a starship by the galactic wizard."

The reporters smiled and scribbled feverishly. Christa worried. Should she have been more aggressive? Should she approach the reporters and make a statement afterward? Had she failed her first big test?

"Hang in there," she heard her husband say. "Relax and have a good time."

Christa walked that evening with Sposato and their colleagues to the National Air and Space Museum. There, on a movie screen five stories high and seven stories wide, they watched *The Dream Is Alive*, a spectacular thirty-seven-minute portrait of space flight that had been produced from footage of three earlier missions. They looked over Judy Resnik's shoulder as she released a communications satellite into orbit. They saw her eating and sleeping in space. They crossed their fingers as Bob Crippen and Dick Scobee tried to maneuver *Challenger* to within arm's reach of an ailing satellite, the shuttle's maneuvering fuel running dangerously low. They witnessed the glory of a launch, a space walk and a landing, and, like many of the others, Christa shed tears of exhilaration.

The next morning, as if *The Dream Is Alive* had not been enough, NASA flew in two astronauts — Resnik and Joseph Allen — to share the wonders of space flight. Allen and Resnik represented the new breed of astronaut: men and women who possessed the right stuff not because they were fearless aviators but because they were scientists, people who knew computers and lasers as well as pilots knew dead-stick landings. Allen held a doctorate in physics from Yale, Resnik a Ph.D. in electrical engineering from the University of Maryland. They each had flown in space, and now here they were to show and tell.

Resnik had attempted her first space flight a year ago to the day. Strapped into the shuttle *Discovery*, her back parallel to the ground, she had listened to the countdown on the headset inside her helmet. Nearby had sat Charles Walker, an engineer for the McDonnell Douglas Corporation and one of NASA's first payload specialists, a nonastronaut conducting experiments for a private contractor. Six seconds before the scheduled lift-off, the first of the shuttle's three main engines had rumbled alive, and the orbiter, as always, had lunged forward a bit, ready for the solid rocket boosters to ignite and propel it skyward. Then, two seconds later, alarms had sounded inside the shuttle. A computer had detected a faulty valve in one of the main engines. The countdown had stopped. The launch had been aborted. A disaster had been averted.

Resnik's mother, Sarah Belfer, had heard the distant rumble turn to silence, and as she waited for rescuers to put out a fire that burned beneath the shuttle, ten stories beneath her daughter, she had rested her forehead on her folded hands as if in prayer. At a press site nearby, John Noble Wilford, an aerospace writer for the *New York Times*, had described the shuttle's brief rumble as "an agonizing groan, perhaps technology's embarrassed sigh." Resnik emerged awhile later, soaking wet and frightened.

Now, as Resnik tried to unravel the mystery of her slide projector in the grand ballroom of the L'Enfant Plaza, she explained that the flight had been postponed two months while engineers repaired and tested the faulty valve. The delay proved that NASA was "very cautious," she said. "We take our time because we only have one chance to do it right."

The teachers listened spellbound as Resnik described the actual mission, first her uneasy march across a 150-foot-high catwalk to the shuttle on launch morning. Her fear of heights made the walk across the steel-grated bridge agonizing enough, Resnik said, but then "you look down on the machine and realize very much that this is no longer just a cold metal composite. It's a living and breathing invention that's about to make a space journey."

On lift-off the rockets fire like giant Roman candles, the shuttle breaks the speed of sound in a few seconds and "there is no doubt in your mind that you're on your way," she said. "It's like a very fast, bumpy train ride. After two minutes the solid rocket boosters will burn out and be cast off, but until then we have quite a ride going. You feel the shock waves, but then your acceleration drops back a little bit as the solid rockets tail off, and there's this giant flash in the rear windows as they [the rockets] come off. About six minutes later, you're in orbit."

Hunched forward, fascinated, Christa listened to Resnik describe the moment a shuttle crew breaks the bonds of gravity and enters a weightless world of perpetual motion. She heard Resnik recall the sixteen sunrises a shuttle crew sees each day, the brilliant bands of blue and red that emerge on the black horizon seconds before the bright fireball of the sun bursts forth. She saw slides of Resnik eating and drinking and playing in space. Christa wanted to go.

"I felt like I was on the middeck," she said. "I felt like I was flying with her. I wasn't ready to come down."

But Allen, who had recently published the best portrayal of a shuttle flight to date, "Entering Space: An Astronaut's Odyssey," brought her home from the imaginary journey. He talked about the sensation of breaking through the Earth's atmosphere, a fiery plunge he later described as returning "from a world that is very quiet and dreamlike to a world with air rushing by the windows, from

157

a world that's pretty much dark to a world that is cloaked in a stream of fire. It's like flying down an endless neon tube, and the light gets brighter and brighter as you come home."

Back home at the L'Enfant Plaza, he told the teachers that he and Resnik were privileged to have been chosen from among 102 astronauts to speak to them.

"We are a very diverse group of people," Allen said of the astronaut corps. "You'd think we fit the stereotype, but nothing is further from the truth. We represent at least two sexes" — Resnik smiled — "a number of religions and many different races. There is only one thing I can think of that we have in common, and that, with no exception, is that each of us has been fortunate enough to get a very good education, primarily because at some time in our careers we have been associated with excellent teachers, and you are *their* representatives."

Resnik wanted the last word. She raised her hand.

"It's a shame every one of you can't be taken, because you're all real winners," she said. "I sure look forward to meeting the lucky person, and with any luck I'll still be on that mission, too."

For teachers who too often had felt neglected, sometimes scorned, by their communities, the adulation was wonderful tonic. Here they were for one fine week in the nation's capital, fussed over by the staff of a fancy hotel, wined and dined by NASA, coddled by the press, praised by the astronauts. Men who usually spent their free periods breaking up scuffles in the school cafeteria now had their shoes shined and their chins raised high. Women who rarely wore dresses into the classroom wore high heels and pearls and makeup. They felt rewarded for once. They were the talk of the town.

"It's like I've won a quiz show and the prizes just keep coming and coming," said Gail Klink, an English teacher from Newark, Ohio.

The next gift was a field trip to Capitol Hill. Just back from his journey as the first sitting senator in space, Jake Garn of Utah had invited the teachers to meet their representatives and to hear him explain the glories of space travel. The flight had reduced the rest of his life to a trifling afterthought, he said. But he warned with a foreboding frown about the rigors of training, particularly "the vomit comet," a NASA jet that plummets from high altitudes to simulate the sensation of weightlessness.

Christa groaned, remembering her child-hood bouts with motion sickness.

Every senator and congressman had been invited to the reception, and Christa, ever curious about the soul of government, watched them parade into the stately senate chamber, pat their candidates on the back and pivot toward the door, smiling. Senator John Glenn stopped by despite his opposition to the space flight participation program as a frivolous and dangerous publicity ploy. So did about thirty other senators and congressmen. Then it struck her that no one, absolutely no one, had arrived from the State of New Hampshire.

After all, her senators and congressmen were Republicans, she thought, and maybe they knew she was a liberal Democrat. Maybe they knew she had campaigned for Jerry Brown for president in 1980. Or maybe they were just too busy. Or didn't care. In any case, here she was, all dressed up and feeling like the only girl at the dance without a date. Finally, Ladwig asked her if she would like to pose with him and Garn. She said she would love to.

It was Tuesday, the eve of her debut before the judges, and as she left Capitol Hill, Christa's calm appearance betrayed her jangled nerves. She slept restlessly that night,

160

tiptoeing through a minefield of conflicting thoughts, a new one each time she woke. Did she really have the right stuff? Would the judges think so? And what if they did? Did she really want to interrupt the life she had so purposefully pursued? And what about Steve and the kids? She was the energy in the family, the compass that kept them on course. Could they *really* get by without her? And all things considered, shouldn't she feel guilty for wanting so dearly to win this little junket?

Christa was waiting when Sposato stepped off the elevator at 5:30 the next morning. Their rubber soles beat in rhythm as they set out for the Vietnam Veterans Memorial, Christa's long hair bouncing on her shoulders, the first sunlight peeking through the black branches of the trees. They ran in silence to the Mall and stopped at the memorial, its scroll of the war dead chiseled in tiny gray letters on polished black marble. The Washington Monument rose to the east; the Lincoln Memorial behind them. They gulped in the cool morning air.

As he caught his breath, Sposato said that his best friend from high school, Thomas Nigrelli, had died in Vietnam a week before he was to come home. Christa had known young men from her hometown and from her husband's military college who had died in the war, but she had known none of them as well as Sposato had known Nigrelli. She helped him search the long black slab for his friend's name, and as they walked, Sposato told her about their teenage years in Westerly, Rhode Island, the good times and the bad, the party they had thrown for Nigrelli before he had left and the party they had planned for his return.

When they found Nigrelli's name, Sposato ran his fingers across the cold stony inscription. He prayed for his friend, and Christa prayed with him. Then they hugged, Sposato's cheeks glistening with sweat, Christa's with tears. The birds sang above them.

On the way back to the hotel, Christa told Sposato about the fears that had plagued her the night before. She admitted missing her family so much that she had already called home three times.

"It's getting pretty costly, you know," she said.

"Costly?" Sposato said, slowing to a walk about a block from the hotel. "Tell me about it. I've called home *fifteen* times already. I could build a new shuttle with the money I've spent."

They talked about the attention they had

received, the unexpected luxuries (when Christa's room-service breakfast had not arrived the day before, the kitchen staff had sent her candy and flowers), and about the seductiveness of celebrity. Christa said she could keep it in perspective, but Sposato had balked.

"It's like a land of Oz," he said. "It's so easy to get overwhelmed by it, to get caught up in the fame. Every night I pray, 'Please don't let me forget what's valuable in my life. Help me remember I'm a teacher. I have a family and a job I love. Help me remember the rest is gravy.' "

"But shouldn't we go for it?" Christa asked, tugged by the will to finish what she had started. "Don't we owe it to ourselves?"

"You have to answer that for yourself," he said, "but when they ask me if I'm ready to sacrifice teaching and time with my family, I'm going to say, 'No, I don't think so.' I'm going to say, 'Do what you want, but you can't take anything away from me. I already feel like I've achieved so much.' "

"Well, then, what are we doing here?" Christa wondered aloud. Then quickly, before Sposato could respond, she said, "I guess we're reaching for the stars."

So on the morning she prayed for a soldier

who died too young, she stood on a cement sidewalk, sweat shining through the wet curls on her forehead, and resolved once and for all to do what she had done for thirty-six years — squeeze the most out of life. She would take the rest as it happened. Her first interview was a few hours away.

At first glance the panel of judges seemed rather unusual. Recruited by Terri Rosenblatt, director of the teacher-in-space project for the Council of Chief State School Officers, there were fourteen men and six women representing government, education, science, medicine, business, sports and the arts. Among them were LeRoy Hay, the 1983 National Teacher of the Year; former astronauts Donald Slayton, Eugene Cernan and Harrison Schmitt; the presidents of Duke University, American University and Vassar; Dr. Robert Jarvik, the inventor of the artificial heart; Wesley Unseld, a former professional basketball player; and actress Pam Dawber, the television girlfriend of Mork, the extraterrestrial from Ork in *Mork and Mindy*.

Some of the choices seemed to make sense, but others, well, even Rosenblatt's boss wondered.

"Pam Dawber? Are you kidding?"

"No, I'm not kidding," Rosenblatt insisted. "Pam Dawber knows what it's like to become famous overnight. She knows what it's like to leave a hotel room for a cup of coffee in the morning and get mobbed before you reach the lobby. She can tell the teachers how winning this thing will change one of their lives, and then she can ask them if they're ready for it."

Each teacher's application and videotape had been delivered to the judges a month earlier. Using both for background information, each judge was to conduct one-on-one interviews with about twenty candidates, then divide into four groups. Each group would nominate four finalists.

Searching for an edge, Christa had studied short biographies of each of her judges: Hortense Canady, president of Delta Sigma Theta, a national honorary society for black women; Konrad Dannenberg, a retired rocket scientist, first for Germany, later for the United States; Sidney Marland, a former U.S. commissioner of education; and Terry Sanford, the president of Duke University. Christa had anticipated their questions and rehearsed her answers, but now, minutes away from the first interview, she was pacing the hotel lobby, a prisoner of self-inflicted pressure. Sposato spotted her.

"Hey, Christa!" he said. "Welcome to *You Bet Your Life!*"

"Charlie, what are you . . . ?"

Suddenly, Sposato was Groucho Marx, darting about the lobby, bent forward, his left arm behind his back, his right hand held high, his fingers flicking an imaginary cigar. He hid behind curtains and careened between the furniture. He insulted an imaginary socialite here, an imaginary bellhop there. He flicked imaginary ashes from one end of the carpet to the other, and now he was rushing toward a tall plant next to Christa. Squinting, he looked up at the plant, flicked his cigar and said in Groucho's gruff voice, "All right, what's the magic word?"

Pressure? What pressure? Sposato's stunt had worked. When the door to the interview room opened, Christa walked in smiling.

The interviews lasted fifteen minutes, and the judges, Marland first, asked a prescribed set of questions. Christa was prepared for most of them: "How active are you as a community leader?" "Do you see yourself as somebody who can relate to people of all ages?" "What would the impact be on your family?"

She answered without hesitation when they asked, "Why do you want to go?"

"I believe I can be a good symbol for all American teachers," she said. "Our profession needs this, and I think I can do the job."

But another question surprised her: "What was your most embarrassing moment?"

She thought for a minute and recalled her first back-to-school night at Concord High School. The parents had come to make sure their children were in good hands. Eager to impress them, Christa had prepared carefully. She had studied her course outlines, memorized the minute every class bell rang, remembered the holiday and vacation schedules. She had spoken for a few minutes and then had offered, with supreme confidence, to answer any question they might have. They had only one — "How do we get to the music room?"

Christa had stammered for a moment, then, blushing, she had admitted she didn't have a clue.

Mercifully, she thought, the teacher-in-space interviews passed more uneventfully than back-to-school night. The judges were kind and sensitive. They put her at ease, even made her feel important. They gave no indication that they preferred her to anyone else, but she was satisfied simply that she had done her best. Now she could relax.

The teachers rode to the White House that afternoon, and Christa sat in the back of the bus with Sposato and John Wells of Puerto Rico. The three of them giggled like schoolchildren as they imagined what it would be like to meet the president. Finally, they burst into laughter, and someone in front of them, acting very much like a teacher, said, "Will you people quiet down?" Taken aback, they were silent for a moment. Then they realized that one of the judges, Harrison Schmitt, a former senator and astronaut, was laughing with them. They continued.

The president was in a humorous mood as well. As the teachers sipped iced tea and lemonade in the East Room, he quoted a British Royal Astronomer who said two years before the first *Sputnik* was launched, "Space travel is utter bilge." Invoking another expert, Reagan said, "The acceleration which must result from the use of rockets inevitably would damage the brain." And he told the teachers, "Your shuttle doesn't blast off for a while yet, so there's still time to back out."

By the way, he said, "for the lucky one who does go up in the shuttle, I have only one assignment: take notes — there will be a quiz after you land."

But he also praised the teachers and their

role in the space program.

"You save our past from being consumed by forgetfulness and our future from being engulfed in ignorance," he said. "When one of you blasts off from Cape Canaveral next January, you will represent hope and opportunity and possibility; you will be the emissary to the next generation of American heroes. And your message will be that our progress, impressive as it is, is only just a beginning; that our achievements, as great as they are, are only a launching pad into the future. Flying up above the atmosphere, you'll be able to truly say that our horizons are not our limits, only new frontiers to be explored."

Someone tried to scale a White House fence while Christa and several other teachers conducted television interviews on the lawn after the reception. Imagining themselves bit players on a Hollywood set, they continued their interviews, pretending not to notice the armed guards who dashed about the grounds. Christa was more amused than frightened, and she laughed louder on the bus ride back to the hotel than she had laughed earlier.

But the party was nearly over for the class of 51-L. At closing ceremonies in the hotel ballroom the next afternoon, William Nixon,

NASA's education chief, gave each of the teachers a NASA pin, a teacher-in-space patch, an award certificate and a "space ambassador's kit" — a duffel bag stuffed with fifty pounds of slides, videotapes, books, more educational material than they could use in a year. Ann Bradley invited them to watch the launch from a VIP viewing area at the Kennedy Space Center. Alan Ladwig paid them a final tribute, and then Terri Rosenblatt explained that ten of them could expect a call at home the next day. The judges were deciding which ten.

The twenty judges — the actress and the basketball star, the astronauts and the college presidents, the inventor and the rocket scientist — had gathered in a conference room NASA had borrowed from the Office of the Controller of Currency in a building next to the hotel. Christa's judges had nominated her first, and none of the others had protested.

"She was outgoing, very enthusiastic," said Konrad Dannenberg, one of the four who interviewed her. "I think we were all very favorably impressed by her spirit, her speaking ability and her ideas about communicating the space experience. The judges who didn't interview her saw those qualities in her application and her videotape. She was chosen quickly."

Others were not. The deliberations lasted for several hours as the judges reviewed applications and videotapes, compared interview notes, debated the merits of substance and style and lobbied for their personal favorites. By the time they had finished, most of the teachers were on their way home.

Christa had stopped on her way out of the hotel lobby to chat with Kate Koch-Laveen, a high school science teacher from Apple Valley, Minnesota, whom she had met at the welcoming banquet six days earlier. Convinced she was no longer a contender for the final ten, Koch-Laveen was conducting an informal poll as the teachers checked out.

"Everyone was either real giddy and excited about their chances or they were a little downcast, knowing they had already played their hand," she said. "Then there was Christa. Christa just seemed so unaffected by the whole amazing week. She really hadn't passed judgment on herself, and she seemed to be at peace with whatever was about to happen. I'll never forget how calm she was."

As they talked, Koch-Laveen handed Christa a business card adorned with a cartoon sketch of a space shuttle. The shuttle's cargo doors were open and a teacher stood at a blackboard before a class of students. Above the shuttle was written, "The Ultimate Field Trip."

Christa smiled and said good-bye. It was Thursday, June 27, 1985.

Come morning Terri Rosenblatt began notifying the ten finalists, feeling every bit like the person who handed out the million-dollar checks on television. She called the winners, congratulated them and told them to keep the news a secret until NASA introduced them at a press conference in Washington on Monday. They were to report to the Johnson Space Center in Houston a week later for physical and psychological testing and a brief encounter with an astronaut's training regimen, most notably a ride on the vomit comet. Those who were still interested in space flight would be interviewed the next week by a committee of senior NASA officials. Then there would be a winner.

Rosenblatt reached Judith Garcia across the river in Alexandria, Virginia, at 7:30 A.M., and by nightfall she had contacted every finalist but one.

Christa had been confused. Believing Rosenblatt would call the finalists on Saturday, not Friday, she had spent Thursday night with Patricia Mangum and returned to

Washington the next day for a few hours of shopping. She had stayed in the capital until early evening and flown to New York, where she had missed a connecting flight to Manchester, New Hampshire. By the time Christa arrived in Manchester and Steve had driven her the twenty minutes to Concord, it was after midnight. Rosenblatt had gone to bed.

Undaunted, however, Rosenblatt had set her alarm for 3:00 A.M.

Christa lurched awake when the phone rang. She nudged Steve, who put on his glasses and groped out of their recently renovated third-floor bedroom and down a flight of stairs to answer it. Christa lay in bed.

"Hello," he said.

"Hi, this is Terri Rosenblatt of the Council of Chief State School Officers. I'm terribly sorry to bother you at three in the morning, but I think you'll find it was worth the wakeup call. May I speak to Christa, please?"

Bleary eyed, Steve tramped back to the bedroom.

"It's for you," he said. "It's someone from NASA. You made the final ten."

Christa shot up in bed and squinted at him, her eyes adjusting to the light. "If you're kidding," she said, half asleep, "you . . . are . . . in . . . big . . . trouble."

173

CHAPTER SIX

Far from the cool shade of her New Hampshire home, Christa rode the sun-seared freeway south of Houston past parched bayous and flaming oil derricks, Tex-Mex restaurants and cowboy shops, horses grazing on the brown, unforgiving flatlands. She turned onto NASA Road 1, thumped over a railroad crossing and came upon a former cattle pasture that for twenty-two years had been the breeding ground of astronauts. Full of hope, she arrived at the Lyndon B. Johnson Space Center to trace the steps of her childhood heroes. How strange the steps turned out to be.

It was Sunday, July 7, 1985, eight days after Terri Rosenblatt had waked her from one dream to start a new one. Christa had kept her promise to Rosenblatt, sharing the news with only her closest relatives and skipping Steve's annual law firm picnic to join the nine other finalists for a press conference in Washington. She had flown home to be toasted by townspeople and mobbed by reporters. ("Come on in," Scott told one of them. "We're getting calls every second.") Then she had packed her coolest summer clothes and set out for Houston — the land of country music, pickup trucks and chicken fried steak — more intent than ever on coming home a winner.

And for once Christa arrived on time.

The teachers were booked for six nights at the Super 8, a budget motel three miles from the space center and a long way from the L'Enfant Plaza. NASA could have afforded more elegant lodging, but the finalists faced a rigorous schedule, they needed no distractions and the Super 8 was one of the last places reporters would search for them. The lucky ones got a room with a view — a six-lane highway that linked Houston and Galveston.

Forget the view, Christa thought. And never mind the heat. Or was it the humidity? Whatever it was, forget it. What mattered were the tests. She and her nine colleagues were on the threshold of the most intensive physical and psychological tests of their lives: pressure tests, strength tests, respiratory tests, claustrophobia tests, psychiatric tests, high-altitude tests, dental tests, blood tests, urine tests, more prodding, poking and probing than they had ever imagined. And then they were to face the clincher — the vomit comet.

NASA had cooked up a good long Texas-style taste of "space flight suitability testing" just so everyone understood that sending people into the sky on the belly of a delicately controlled bomb was downright serious business.

"Some of them had a romanticized view of what they were getting into," Rosenblatt said. "Pretty soon they were grappling with the fact that this was a major commitment, something that could change their lives."

Not everyone would yearn for space flight as eagerly at the end of the week as they did now, Christa suspected, and she hoped she wasn't one of them. As NASA officials briefed the ten of them in a cramped motel conference room, she wondered about the other nine. She hardly knew them. She knew from reading their résumés that the youngest was thirty-three, the oldest forty-five, and that they had an average of fourteen years of teaching experience. She knew that only two of them taught math or science, and that only one of them taught in a big city. Beyond that, she knew she was no more qualified to ride in space than any of them, at least not on paper.

Take Richard Methia, she thought.

A high school English teacher, Methia had spent his life in New Bedford, Massachusetts, a city with a glorious past and a dismal present. He had been married and raised a daughter there, and he had seen New Bedford decline from a thriving seaport, once the whaling capital of the world, to a city strangled by unemployment, poverty and crime. Trying to save it, he had mediated disputes between street gangs and the police, established teenage job programs and founded the city's first halfway house for troubled youth. As a union leader in the 1970s, he had gone to jail twice to try to improve conditions for the city's teachers.

Now he was trying to heal the city's ugliest wound. Four men had raped a woman on a pool table at Big Dan's bar as patrons cheered. Bearing the nation's anger, the media had descended in all its fury, led by the Cable News Network, which broadcast the trial detail by unseemly detail — live from New Bedford. A year had passed since the men went to prison, and the city had yet to recover.

"I want the [shuttle] experience for my hometown," Methia wrote to NASA. "I want the spotlight to shine on my community — a city too ready to believe its glory lies buried in its past."

Not only was he a spirited public servant, he was also far and away the most eloquent of the finalists. He quoted Shakespeare in his videotaped interview and began his essay on why he wanted to be the first private citizen in space with a poetic flourish: "In *Inherit the Wind,* we are warned that when man learns to fly, the birds lose their sense of wonder and the clouds begin to smell of gasoline. I want to prove that isn't true."

He had written poetry, short stories and prize-winning plays, but nothing Methia had written intimidated Christa and the other finalists more than his essay that appeared in *Newsweek* the week they arrived in Houston. The article, "Riding the Blackboard Shuttle," attacked the National Education Association for criticizing the teacher-in-space program. Millions read it, including members of NASA's final selection committee.

"It is graceless to demean [the program] and shortsighted to ignore it," he wrote, "for across the nation this simple gesture can work much good. . . . If it sparks idealism in even a few young minds, if it frees from the dreary sameness of commonplace ideas even a few older minds, if it entices adults to believe once more in their own dreams, if it reminds us that progress is the conquest of the ordinary by the marvelous, then this single adventure for one of us will feed the hungry human spirit in all of us."

Powerful stuff, Christa thought. And, worse, he spoke as eloquently as he wrote. She had no way of knowing exactly what NASA wanted, but in the world of education at least, she knew Dick Methia oozed the right stuff. He had even shaved his beard in the quest for a shuttle seat.

None of the other finalists had written for *Newsweek* or gone to jail for their profession. They possessed simpler gifts.

Robert Foerster, a sixth-grade math and science teacher from West Lafayette, Indiana, had designed software programs and helped improve the use of computers in the nation's schools. He had rosy cheeks and an easy smile, a wholesome, all-American look that would play from Peoria to the Florida panhandle. His shuttle project — to demonstrate the properties of weightlessness to the common man by using marshmallows and gumdrops, among other items — was appealing. His style was clean and simple.

But David Marquart was NASA's man if it wanted someone who waved the flag as well as he taught. In twenty years of teaching business and computer science in Boise, Idaho,

Marquart had served as a director of the U.S. Air Force Radio System in Idaho, as a training officer for the Civil Air Patrol and as the producer of a promotional film for a local Army ROTC unit. He was a Boy Scout leader with John Denver's looks and the soul of Uncle Sam.

"I want to fly in space as the first private U.S. citizen to show the world that only in America can an individual achieve this recognition without being directly involved in the military," he wrote. "Only through the good will and graciousness of our President has a teacher been considered as the first layperson to share this experience."

No finalist, however, had stronger ties to the military than Michael Metcalf, a government and geography teacher from Greensboro, Vermont. A former Air Force pilot, Metcalf wanted to fly in space because "to anyone who has 'slipped the surly bonds of Earth,' such flights of fancy are indeed appealing. The thrill of supersonic flight into the deep blue of the stratosphere, the heady surge that fills one with awe at the beauty of a world most men never know, all of this impels [me]."

After a military tour in Europe, Metcalf had settled in Greensboro (population: 675), where he served on the board of selectmen and the zoning board. He owned and operated a small solid-waste management business, wore a tiny American flag on his lapel and described nearly everything that appealed to him as "super."

"This is a super crew of people," he said of the finalists. "An absolutely super group of men and women."

Of his chance to fly on the shuttle, however, he said, "I'm as tickled as a kid with a green egg!"

Each of the male finalists was attractive, articulate and personable, but if any of them was chosen, Christa thought, even Methia, she would have gone home angry.

"Look back at all the teachers you've had in your life," she said. "Who comes to mind? Historically, teaching and nursing are among the few professions that have not been dominated by men. Each of the male finalists is more than qualified, but if you're going to choose someone to represent teachers as a whole, I think you should be truly representative. You should choose a woman."

NASA had six to choose from, and two of them — Kathy Beres and Barbara Morgan — impressed Christa the most.

Beres had been exploring since she was nine

years old and named her turtle "Sputnik" for the first satellite to enter space. She had climbed the Andes and the Himalayas, led a rope team on a three-week journey across the northern glaciers and sailed a thirty-one-foot boat across the Atlantic. She had done field work in Iceland, East Africa and the Galápagos Islands, and she had begun planning for an expedition to Antarctica.

But first she wanted to travel through "the ocean of space."

"The zest for life and adventure and a desire to do the daring have always been a part of me," she wrote. "Space at our fingertips offers untold promise. It lures us ever onward."

Between adventures, Beres taught high school science in Baltimore, her lifelong home. But she sounded as if she were from southern California. She said "rilly" instead of "really," "to the max" instead of "extremely," and when she had just remembered something, she said, "Isn't the brain an incredible disc? Sometimes we recall things we never knew we had stored." Tanned and blond, she even looked like she lived in southern California.

Her enthusiasm was her greatest strength and like Christa she encouraged her students to squeeze the most out of life.

"Enjoy it," she told them. "This is no dress rehearsal."

Her shuttle project was her only clear weakness. Recognizing NASA's desire for a private citizen's perspective of space flight, most of the finalists had proposed video or written documentaries, live television lessons or photo journals, nothing as ponderous as Beres's "detection of material at the Lagrangean points," a study of the gravitational field in the earth-moon system.

The judges in Washington, some of whom knew far more about rating points than Lagrangean points, had voted for her despite her project. But how would the beer-drinking truck driver in Orlando understand it? Never mind the eighty-three-year-old widow in Queens. Or Johnny Carson.

Barbara Morgan spoke a simpler language. The daughter of a prominent cardiologist in Fresno, California, Morgan had not allowed the trappings of success to spoil her childhood sense of wonder. From her home on the shore of Payette Lake, a trout-fishing paradise a mile above sea level in McCall, Idaho, she still gazed at the stars through the telescope she had used as a child. She lived by the same spirit that prompted her at the age of seven to ask why a monkey had gotten to enter space instead of her, and at the age of eight to announce that one day she would ride with the Royal Canadian Mounted Police. She saw the world through a child's eyes.

Unlike Beres, who seemed to have consulted science journals for her shuttle project, Morgan had consulted her second-grade students. She was searching for the answers to their questions:

"What time is it in space?"

"Would a kid be as strong as a grown-up in space?"

"Do you dream in space?"

"Do the stars look the same?"

"Can you take a deep breath and blow yourself across the room?"

"Does a flame burn upwards or downwards?"

"What happens when you bleed in space?"

Curiosity compelled her.

"The great thing about being on Earth," she said, "is having the opportunity to live and learn about as much as you can."

After graduating Phi Beta Kappa from Stanford, where she met her future husband, Clay, she had taught remedial reading to children of the Salish and Kootenai tribes on the Flathead Indian Reservation in Montana. A year later, she married Clay, settled on Payette

184

Lake and started teaching second grade in McCall. Clay wrote novels in the winter and jumped out of planes in the summer to fight fires for the U.S. Forest Service. Three years later, they traveled to Ecuador, Barbara to teach, Clay to research a book.

"You can't really know about something unless you get a little bit of it on you," she said. Now she wanted "to get some stardust on me."

Morgan had gotten a little bit of a lot of things on her in the week before she arrived in Houston. She had climbed a mountain, sailed on Payette Lake, played her flute with friends from the McCall Chamber Orchestra and ridden her bicycle into town to eat rum tarts at the Danish Mill coffeehouse. She arrived at the space center with a camera dangling from her neck, her eyes aglitter and her dark-brown, shoulder-length hair pulled back on one side with a clip. She looked like a child on her first day at Disneyland, Christa thought. She looked like she belonged on the shuttle.

Of course, three other women believed they belonged as well.

After twenty-one years of teaching foreign languages in Alexandria, Virginia, Judy Garcia wanted to teach from the blackboard shuttle. She hoped to beam back lessons in three languages — English, French and Spanish — and intended to enhance the mission's international flavor by naming it the "Liberty Flight" in honor of the Statue of Liberty's one hundredth birthday. A spry, inquisitive woman with an engaging smile, Garcia's only problem was that she appeared uncomfortable before the cameras. NASA's teachernaut needed charisma, and at first glance Garcia seemed to need more of it.

Peggy Lathlaen was more animated. A teacher of gifted children in Friendswood, Texas, ten minutes from the space center, Lathlaen projected an appealing air of innocence. In conversation, she sat on the edge of her chair, her brown eyes flashing, her hands cutting circles in the air, speaking as passionately about the weather as she did about her rather curious shuttle project — a study of the reproductive systems of houseflies. She had settled on the project after brainstorming with her class of gifted fifth graders.

"As a potential future mother, I think it is important to follow up the findings on reproductive anomalies in space-traveling flies," she wrote.

The pressure of the competition had begun to nag her, however, and she seemed to have more potential as a future mother than a space flight participant.

Finally, there was Niki Wenger, the Walter Mitty of the final ten. A teacher of gifted junior high school students in Parkersburg, West Virginia, Wenger had graduated from college in 1962, raised three sons and started teaching only five and a half years before she entered the space contest. Two of her sons were studying to be astronauts, and now, at the age of forty-five, she had a chance to beat them into space.

"For as long as I can remember, I have been mesmerized by the night sky," she wrote in an impassioned plea. "I can spend hours searching for falling stars, marveling at the Milky Way, studying the planets and the moon, getting lost in the constellations.... How can I adequately convey to you why I want to go? I feel an intense yearning to be out there, to know for myself that vast wonderous splendor."

But she had also missed NASA's point. Instead of using the flight as a teaching tool, she intended to conduct research, hoping, among other things, to determine whether space sickness could be prevented with meditation, biofeedback, sensory deprivation or self-hypnosis. The project would never fly.

No, Christa thought, not all of her colleagues

would yearn for a shuttle ride as much at the end of the week as they did now, nor would NASA continue to be equally impressed by each of them. She wished she could see the future.

When the briefing broke up, she bade the group good night and returned to her room. Her first test was only hours away and the last thing she needed was to miss it. She asked for a 5:30 A.M. wakeup call and went to bed while the others lingered downstairs.

"Christa was rather quiet and subdued all week," Bob Foerster said. "She knew what she wanted to do and she was very determined about it."

She woke with a start the next morning. She had missed her first appointment. She was sure of it. The wakeup call had never come, and she had slept away the chance of a lifetime. She knew it. Hurling off her covers, Christa groped her way to the bathroom, started the shower and rushed back to her night table to look at her watch to see just how late she was. Not that it really mattered. Why should NASA trust her to show up for a shuttle launch if she couldn't make it to a doctor's appointment on time? She was finished. She knew it. Then she picked up her watch. It was 2:00 A.M.

In one shivering moment the adrenaline stopped vibrating through her veins and she slumped onto the bed, relieved and at the same time convinced that winning this contest meant every bit as much to her as she had suspected. She wondered if her blood pressure would drop by morning.

When the phone finally rang, Christa bolted awake, showered and stepped into the hot Houston haze, her empty stomach tingling with anticipation. Within minutes a government van had whisked the ten of them to NASA's high-tech holy land, a sixteen-hundred-acre compound of bright green grass, man-made ponds and a hundred cream-colored concrete buildings that stood on streets named Gamma Link and Beta Link. Satellite dishes flourished like wild flowers. Miles of chain-link fence kept intruders at bay.

In sharp contrast to the hurly-burly of the shopping malls and the fast-food restaurants on NASA Road 1, the space center had the simple, symmetrical look of a suburban industrial park. Most of its thirteen thousand workers were scientists, engineers and technicians, men who wore white shirts with plastic pencil holders in their pockets; women who were polite, brisk and businesslike. Everyone spoke a cold, abbreviated language that required a 305-page dictionary to decipher.

If you asked for a TACO at the Johnson Space Center, you got "a test and checkout operation." TLC had nothing to do with affection — it meant "telecommand" — and LSD was "low speed data." But if someone mentioned MT, you needed more than a dictionary. You needed a clue. Did they mean magnetic tape or a master timer? A master tool or a mechanical technician? Maybe mean time, mission time or mountain time? Maximum torque, perhaps? Or mission trajectory? It was all very confusing.

Now the TIS (teacher-in-space) candidates who hoped to be aboard STS (shuttle transportation system) flight 51-L upon L/O (liftoff) from launch pad 39B at KSC (Kennedy Space Center) were at JSC to see the MDs. They blew in like a fresh breeze.

"NASA?" Morgan said. "What is that: the National Association for the Suggestion of Acronyms?"

Christa started the day at Building 8, where she donated 7 vials of blood, answered 668 questions about her medical history and spent 3 hours as a lab rat, submitting to X rays, an eye test, a dental checkup, Dr. Hein's dreaded proctology exam, a pulmonary-function study and a series of procedures she had never heard of, including a musculoskeletal test in which technicians measured every inch of her body. She felt as if she were competing in a prize fight and a beauty contest on the same day.

"My gosh," she said, "they even know the height of my bellybutton."

The doctors had forbidden her to eat for the previous fifteen hours, so she had a Texas-size appetite by lunch. The technicians warned her not to try to catch up on lost calories, though, because next came the treadmill, NASA's space-age running machine.

Bathed in the white light of a white room, Christa smiled quizzically as a dozen electrodes were attached to her torso, a blood-pressure gauge to her arm and breathing apparatus to her head. She wore a green hospital smock, white ankle socks and jogging shoes, and as she stepped onto the inclined conveyor belt she wished for the life of her she could trade it for a sidewalk in Concord. She wondered what Alan Shepard would have thought of all this, and then — it caught her by surprise — the machine started to roll.

Don't fall off, she told herself. Whatever you do, don't fall.

Moving slowly at first, the treadmill gained speed until she had advanced from an easy stroll to a trot to a vigorous jog. Any second now, she thought, she would begin flailing like some cartoon character trapped on a conveyor belt. She tried to grip the railings, but the technicians forbade it. Instead, Christa draped her arms over them and ran like Frankenstein until the machine slowed mercifully to a stop thirteen minutes later. She had survived the first battery of tests.

Christa faced the press the next morning — the networks, the giant dailies, the magazines, the hometown papers, the wire services and the cameras, lots of cameras. Even in its post-Watergate cynicism, the media still had room for a sacred cow, and NASA was it. Instead of pressing reporters to probe the nuts and bolts of the space program, editors, sensing a resurgence of heroism in the eighties, preferred stories that portrayed the people who explored the heavens as larger than life. No one was fooled by NASA's plan to send a teacher into space. The idea was to light a fire under millions of Americans who had begun to consider shuttle flights routine, and the media were ready to strike the match.

Christa, of course, was eager to cooperate. She had a nervous giggle and the gee-whiz bounce of a camp counselor, but she was more comfortable than ever among faceless photographers and other inquisitive strangers. She rushed from one interview to another, posing before a mock shuttle the astronauts used for training.

"Oh, I love it," she gushed, looking up at it. "It makes you want to hop on board."

When television reporters asked her what she thought of the space center, she smiled and told them she felt like a kid in a candy store. When newspaper reporters tempted her with tougher questions, she took them on.

"Couldn't the six hundred fifty thousand dollars it will cost to select, train and send a teacher into space be better spent in the schools?" one of them asked.

"You know, I think this is the best public relations gift to the schools that NASA could ever make," she said. "It's a bargain when you think of the students and teachers we'll reach. There's no other way you could get so much done for that kind of money."

"What about teachers' salaries?" another one asked her. "Are they competitive?"

"Well, I make more than twenty thousand dollars, but I think after twelve years of teaching I should be making more than thirty thousand," she said. "I have a sister ten years younger than me in computers who went over thirty thousand dollars several years ago.

That's great for her, but are teachers' salaries competitive? No. A person looking at careers today might find themselves locked out of education."

Christa explained that an English teacher at Concord High who had recently been named the state's teacher of the year earned $23,000 after forty years on the job. He was plagued, like teachers across the country, by the myth of immediate results.

"Teachers don't put out fires or arrest people," she said. "We deal with minds. All we can hand people at the end of the year is hope for the future. How do you translate that into dollars?"

Right there in Houston, she said, the school district was so desperate for teachers that it planned to hire out-of-work professionals and housewives who had no education degrees. The nation had lost respect for its teachers, and she wanted the teacher-in-space program to help restore that respect.

When a reporter asked her why she wanted to go into space, Christa said if she had lived in an earlier time she would have ridden on the *Mayflower,* Conestoga wagons or in early airplanes. She talked about her journal, about how her perspective as an ordinary person would "demystify" the space program and about her vision of the world as a global village, of one people living together. She felt a special affinity for the common people, Christa said, because "the common people rarely start wars."

"Do you have any fear of the shuttle?" another wanted to know.

"Not yet," she said, smiling, "but maybe once the flier's picked . . . "

"What about your family? How do they feel?"

She said her husband wanted to be the first lawyer in space, her son was tickled that one of his stuffed frogs had a chance to fly and her daughter was still confused. Then she excused herself. It was time for the claustrophobia test.

"I have to tell you," Christa said, looking over her shoulder, "I have some real misgivings about going through with this one."

The object of her anxiety — a personal rescue sphere — lay in Building 37 on an observation room's cold linoleum floor. NASA employees called it the PRS, a thick, inflatable nylon ball that was thirty-six inches in diameter, linked to an oxygen supply and being developed for emergencies in space. Only two space suits fit on board the shuttle, and in case of an emergency, NASA expected the commander and the pilot to slip into the space suits, zip the rest of the crew into personal rescue spheres and carry them weightlessly to another shuttle or an orbiting space station. That was the idea, anyway.

Until then the PRS was a test lab, a tiny dark bubble in which the teachers were fitted with transmitters and electrodes and zipped in, one by one, to determine how they might react aboard the shuttle, where the total living space — kitchen, bathroom, sleeping area, living room and recreation area — measured ten by thirteen, no bigger than Christa's kitchen and no place for a claustrophobe. No one told them when they would be released.

Morgan survived by humming Bach's Brandenburg Concerto No. 4.

"The magic of Bach played in my head and I relaxed and enjoyed myself," she said. "My EEG readouts probably looked like sheet music."

Foerster felt "embryonic" in the bubble. "It was dark and warm," he said, "and I started fantasizing that I was lost in space."

Methia, who was mildly claustrophobic, entered the ball in terror, but even he managed to keep his composure.

Then it was Christa's turn. As she curled her five-foot-six frame into the space-age beach ball, she tried to imagine herself crawling into a pup tent in the mountains of New Hampshire. She smiled meekly as a technician zipped her in, switched off the light in the observation room and shut the door.

"I thought I was going to start yelling and clawing to get out," she said.

Then she laid back and folded her arms across her stomach. Christa was alone in the blackness. The air was cool and comforting. She imagined herself floating in space. It was the most peace she had had all week, and when the technician unzipped her fifteen minutes later, she asked if she could take the ball home with her.

"When things start to get crazy," she said, "I can just set the timer and tell the kids, 'Okay, Mom's going into the sphere now.'"

After an afternoon of high-tech strength tests, Christa joined the rest of the teachers, Rosenblatt and Alan Ladwig for dinner at one of the restaurants on the neon strip near the space center. The first battery of tests was behind them and the ride on the KC-135 was still two days away, so the mood was festive. The conversation was so loud that Christa and Kathy Beres, who sat across from each other, had to shout to be heard.

"Let me show you my kids!" Christa said as she reached for her wallet.

She showed Beres a picture of Caroline and then one of Scott at his First Holy Communion. Beres mentioned that she was also Catholic and that one of her nephews had recently celebrated his first communion. She said the boy's parents had engraved his initials on the chalice they had bought for him to use, so he could save it and use it again at his wedding mass. Christa liked the idea. She said the children in her church celebrated their first communion as a group, however, and that the priest might not allow Caroline to use her own chalice. But she said she would ask him.

The next morning the teachers went back to school. For five and a half hours they attended crash courses on the perils of space flight — decompression sickness, hyperventilation, respiratory and circulatory problems, spatial disorientation, hypoxia and more. They scribbled notes, strafed the instructors with questions and volunteered for all the demonstrations. Even Christa, who had never completely conquered her childhood motion sickness, wanted so badly to impress NASA's instructors that she volunteered for a ride on the spatial disorientation chair, a sort of earthbound vomit comet.

The purpose of the automated chair was to help astronauts prepare for the first dizzying moments of weightlessness in space. Christa climbed in, fastened her safety belt and closed her eyes. The chair started to spin, quickly gaining speed, and within seconds it spun so fast that she thought, seat belt or no seat belt, she would be hurled halfway across Texas. Then her inner ear started to mutiny and she lost her equilibrium.

"Are you still spinning?" the instructor asked her after a while.

"No?" she said, hoping for her stomach's sake that she wasn't.

"Take a look," he said.

Sure enough, the room spun faster than the carnival rides that had turned her stomach twenty-five years earlier. She suddenly felt ill. She wanted to get off. She wanted to lie down. She wanted to go home. She wanted no part of the vomit comet.

Which, of course, all this was leading up to. And now it was time for the altitude chamber — a horrible green cylinder that turned Christa's fingers blue and stole her sense of reason. An instructor had drilled the finalists for several hours about the chamber. An emergency aboard the KC-135 or the space shuttle could lead to hypoxia, he told the group, an oxygen deficiency in the body tissue that was characterized by such symptoms as blurred vision, dizziness, a tingling sensation, incoordination, hyperventilation and personality changes that included belligerence and . . .

"Euphoria," he said, "like the feeling you get when you drink three martinis in five minutes."

Each person experienced different symptoms, but for that person the same symptoms would always recur. The purpose of the altitude chamber was for them to learn their symptoms and how to respond to them.

"When you recognize the symptoms, you must get oxygen immediately," the instructor said. "People die from hypoxia."

The chamber was large enough for ten teachers, four instructors and Ladwig, who had decided to ride the KC-135 as well. Wearing oxygen masks, they sat across from each other on steel benches, breathing pure oxygen for twenty-five minutes to cleanse their bodies of nitrogen that could create deadly bubbles in their blood. They used the time to adapt to the masks, which required them to forcibly exhale instead of inhale, reversing their natural breathing patterns. Some of them adapted better than others.

Peggy Lathlaen's mask felt out of place, but she was too nervous to ask for help. Once the steel doors had been locked and the oxygen had been sucked out of the sealed chamber, the occupants climbed to a simulated height of 6,000 feet and Lathlaen looked at Christa with panic in her eyes. Christa had panicked enough for one day. Her eyes were peaceful now, Lathlaen said, and they soothed her. Her mask was fine.

From 6,000 feet, the teachers simulated a plunge to 1,000 feet to be sure they could clear their ears. Then they soared to 35,000 feet and descended to 28,000, where the test began. The occupants across from Christa removed their masks first. Directly opposite her sat Dick Methia.

Still haunted by his "personal demon" — a mild case of claustrophobia — Methia feared the altitude chamber more than the personal rescue sphere, he said, because he "didn't fully understand what was going to happen, and once I realized there was no escaping, the idea of physical danger became even more acute."

At 28,000 feet, the oxygen was so thin that the symptoms of hypoxia appeared quickly. Methia, like the other teachers, held a sheet of paper on which he was to record his symp-

201

toms as he answered a series of questions: name, address, telephone number and several basic math problems. As the other teachers recognized their symptoms, they replaced their masks and replenished the oxygen in their bodies. Methia did not.

His personality had changed. The teacher who had tried to bring peace to the streets of New Bedford began acting like a waterfront thug. The instructors noticed his fingers turning an icy blue from a lack of oxygen and told him to replace his mask. He refused. They *ordered* him to replace his mask and again he refused. Then they tried to replace it for him but he resisted.

"I don't need your help," he said. "I'll do it. I'll do it."

Finally, they overpowered him and replaced it.

"I became extremely arrogant," he said later. "Had I been a pilot, I would have arrogantly crashed my plane."

Once Methia had come to his senses, the teachers on the other side of the chamber removed their masks. Christa began to answer the questions, and soon her vision began to blur. As she subtracted 10 from 63, she noticed her fingers had started to turn blue. Then she noticed her answer: 59.

"I realized then that I probably needed some more oxygen," she said.

No one else reacted as Methia had, and the teachers were dismissed. But some of them, like Christa, still felt uneasy about the KC-135; others suddenly felt more confident.

"I learned more in that chamber than just the symptoms of hypoxia," Morgan wrote later in her hometown paper, the *Central Idaho Star-News*. "I learned to trust the equipment. I knew I could trust the instructors. There is no way that NASA would put ten innocent teachers in danger."

The teachers moved from the altitude chamber into a decompression chamber, where they experienced a sudden decompression from 8,000 feet to 30,000 feet. It began with a loud bang, and the chamber quickly filled with fog, just as it would, the instructor said, if the cabin of their aircraft suddenly lost pressure. No one had a problem, and everyone left smiling.

By the time they returned to the Super 8, they had less than a hour to shower and dress for a reception at the mayor's mansion in Peggy Lathlaen's hometown of Friendswood.

"So much is happening so fast," Christa told a reporter for the *Concord Monitor*, her hometown paper, while her hair dried. "I feel like I'm under a waterfall. The water keeps falling and there's no sign it's going to stop."

Worse, she missed sharing it all with Steve. "He's so supportive and so envious of what I'm doing," she said. "There's so much I've wanted to tell him right away."

She remembered on the way to the mayor's that she had forgotten to call home to say good night to Scott and Caroline. It nagged at her during the reception, and she rushed to a pay phone when the teachers stopped for pizza on their way back to the motel. It was too late to talk to the children, but while the others ate, Christa talked to Steve until her pizza was cold.

The next morning, without explanation, the doctors asked her for more blood and urine samples. Something was wrong, she thought, terribly wrong. After all this — the treadmill, the spatial disorientation chair, the altitude chamber, six months of competition — she was about to lose her chance to fly on a minor medical technicality. As she joined the group for a tour of the space center, Christa tried to conceal her disappointment.

She kept smiling as she tried on an Apollo space helmet in the museum, wriggled into the mock space shuttle at Building 9A and visited Mission Control, where a large digital clock recorded the countdown to the next day's scheduled launch of America's fiftieth manned space flight, the nineteenth by a space shuttle, the eighth by the shuttle *Challenger*. Mission Control reminded her of Frank "Shorty" Powers, the baritone public affairs officer who had led a national television audience in the countdown to America's first manned space launch twenty-four years earlier. She remembered sitting in the cafeteria at Lincoln Junior High, scribbling notes as the rocket soared, and now she shook her head as she considered how far she and the space program had come, and how much farther she hoped to go.

The teachers ate lunch in the space center cafeteria with several NASA officials, and Christa sat near astronauts Bob Crippen, Dale Gardner and Judy Resnik. She mentioned her anxiety about the vomit comet, and Crippen told her to forget it. Forget that 70 percent of the passengers, including Senator Garn, had gotten sick. He told her the flight would beat her only if she let it. Eat a good breakfast, he told her, and show no fear.

Resnik, who planned to join the teachers on the flight, told Christa she had nothing to worry about.

"You know how it is," Resnik said softly. "Only the men get sick."

As the teachers left the cafeteria, an army of autograph seekers mobbed them, causing Christa to balk for a moment. In twelve years of teaching she had signed her share of hall passes, but this seemed, well . . .

"It seemed so silly at first, maybe because I never collected autographs," she said. "Then I saw what they were doing. They were grabbing all ten of us to make sure they had the autograph of the first teacher in space. They realized one of us was going to make history. They wanted to be a part of it."

The three veteran astronauts walked past the swelling crowd, and no one asked for their autographs. No one seemed to notice them.

A few minutes later Christa sat on a cool granite bench in the shade and talked with a reporter about the hints of fame all around her. In Houston there were autograph seekers, the national press, entrepreneurs trying to sell the finalists on everything from books about the first private citizen in space to teacher-in-space sweat shirts. At home there were lecture offers, bouquets, a commendation from the governor (" . . . and whereas all of the citizens of New Hampshire are proud and honored by the selection of Christa

McAuliffe as a finalist . . . "), cards from people she hadn't seen in twenty years, calls from people she had never seen in her life. She had never known such celebrity. She talked about the possibilities.

"Wouldn't it be wonderful for a history teacher to make history?" she said. "What more could a history teacher ask for?"

Christa had never witnessed a shuttle launch, and the next afternoon, after their ride on the vomit comet, the teachers were to watch *Challenger* lift off from the VIP viewing area at Mission Control.

"I don't care how sick I get on the KC-135," she said. "I'm going to see that launch if they have to prop me against a wall."

She glanced at her watch after a while and jumped off the bench. She had to run, she told the reporter. The psychiatrist was waiting.

No one in the history of U.S. manned space flight had left the launch pad without clearance from Dr. Terrence McGuire, NASA's consulting psychiatrist. The agency's flight safety record was perfect, but McGuire needed to be sure that everyone who boarded the shuttle, particularly the first space flight participant, could tolerate high-level stress.

How high?

"Threat to life, a real emergency," McGuire told Neil Chesanow of *New Woman* magazine. "Everything is going fine, and then suppose — well, a seal breaks. And suddenly you're in big trouble. In a situation like that, you need clarity of mind and the ability of move *now*."

NASA had never shot an ordinary person into space, so in the teachers' case McGuire also needed to know whether they had the ability to get along with the crew for four months of training and six days in orbit.

"Astronauts are extremely intelligent, highly trained individuals," he said. "Most have their doctorates in some hard science. They are a very perfectionist, very adventuresome group of people. They expect good performance, good judgment, logical thinking, dedication to the mission, and reliability. You need a person who can fit in."

He judged each of the teachers by a one-hour written exam and a two-hour personal interview. Most of the written questions were multiple choice, and Christa had a hard time settling on a single answer. Asked what kind of animal her friends saw her as — (a) fox, (b) beaver, (c) cat, (d) puppy, (e) owl or (f) lion — she answered (c) because she was independent like a cat and (b) because she was as busy as a beaver.

"I always like to qualify my answers a little," she said.

She faced McGuire after the written test, and to calm herself she used a technique Rosenblatt had shown the finalists before they were introduced to the press in Washington. Count to three aloud, Rosenblatt had said, and then exhale hard, making a grunting sound at the same time. It had worked in Washington, so Christa tried it again.

"One-two-three . . . hunh!"

She walked in calmly and remained calm.

"I tried very hard not to give out any strange body language," she said. "I tried to make sure I smiled and looked pleasant the whole time. It was a long two hours."

McGuire would not reveal details of the interview because of the confidentiality of the doctor-patient relationship, but he said each of the teachers had passed his test and that he had ranked them in order of their psychological suitability as a space flight participant. Christa had ranked the highest.

"In my opinion, she was the most broad-based, best-balanced person of the ten," he said. "A lot of people just don't see themselves as being okay these days. Someone like Christa has a more objective view of who she is and what she's about. That means understanding and accepting her vulnerabilities as well as her strengths. That doesn't mean she thinks she's perfect, that she isn't changing or doesn't want to change. But she has a good idea of who she is, and what she sees is pretty good. That's very unusual today. I know this doesn't sound very scientific, but I think she's neat."

She learned the results of her additional lab tests later that afternoon. Her blood sample revealed a slight deficiency of phosphorus, not low enough to disqualify her, and the second urine sample had been necessary only because the first one had been contaminated. Relieved, she joined the rest of the finalists for supper.

It was the eve of the KC-135 flight, and when the owner of the restaurant learned who his guests were and what they were about to do, he delivered each of them a complimentary cocktail — a kamikaze. They laughed, toasted each other and stopped for ice cream cones at a Baskin-Robbins on the ride back to the motel. Still, despite the advice of Crippen and Resnik, Christa worried about the ride, and she called Steve for encouragement. He told her everything would be fine. Relax, he said, and call me as soon as you land.

She slept well that night, much better, at

least, than her first night in Houston. She rose at daybreak, ready for breakfast but not quite sure she wanted to eat it. The flight was three hours away, and she knew the press would be waiting on the runway to dutifully report her condition before and after.

"NASA told us it was supposed to be fun," she said, "but with that in the back of my head it was hard to think about having a good time."

She followed Crippen's advice, though, and ate well — scrambled eggs, English muffins, tea and orange juice — before riding to nearby Ellington Air Force Base for the final briefing and the flight. The KC-135 would produce weightlessness by flying in a parabolic arc shaped like the big dipper of a roller coaster. After climbing six miles high at nearly the speed of sound, the pilot would shift into a power dive, plummeting at a 45-degree angle that created twice the force of gravity, similar to the force shuttle riders felt upon lift-off and landing. At the bottom of the dive, he would pull up the nose and coast for thirty to forty seconds. The jet would not be supported by gravity, and the occupants would not be supported by the plane. They would be weightless, the closest nine of them would come to the feeling of space flight.

After an instructor had reviewed the emergency escape procedures, including the parachutes, he reminded the teachers about the frequency of illness aboard the jet, handing each of them an air sickness bag.

"But please don't worry about it," he said. "Have a good time."

They each received a dose of Scopedex, a combination of scopolamine, to help prevent air sickness, and dexedrine, to prevent the drowsiness caused by the scopolamine. Christa and Beres hesitated before taking it. They knew dexedrine was a potentially dangerous amphetamine.

"Did you ever take anything like this before?" Beres asked.

"No," Christa said. "Did you?"

"No."

"Do you think we should?"

"Well, it was prescribed by the flight physician," Beres said. "It's not like we're dealing with junkies here."

Christa agreed and they set out for the water fountain, still a bit uneasy. Staring at each other, they rolled the capsules in their hands.

"It was like the tragedy scene in *Romeo and Juliet*," Beres said. "We were sure we were about to poison ourselves."

Christa finally said "Here goes," and swallowed hers, Beres swallowing hers just as Judy Resnik came up behind them.

A complex woman, Resnik had spent her life excelling in a male-dominated world. She had been a straight A student and the only girl in the mathematics club at Firestone High School in Akron, Ohio. She had decided against pursuing a career as a concert pianist, instead studying math and science at Carnegie-Mellon University in Pittsburgh, where she had graduated near the top of a class dominated by men. In 1978, after earning her doctorate in electrical engineering at the University of Maryland, she and five other women, including Sally Ride, had broken the exclusively male ranks of NASA's astronaut corps. She was proud and independent, and she credited no one but herself for her success. She bristled when people described her as "the second woman in space" or "the first Jewish astronaut."

"I am an astronaut," she said. "Not a woman astronaut. Not a Jewish astronaut. An astronaut."

She had another side as well. As a sorority sister at Carnegie-Mellon, she had been elected the runner-up to the homecoming queen. People at the space center knew her as JR. She worked hard, she played hard and she was intensely private about almost everything but her unrequited love for the actor Tom Selleck. Resnik's crew mates on her first space flight had surprised her by taping a poster of Selleck to the inside of the shuttle's bathroom door.

"Excuse me," it said on her coffee cup, "I'm saving myself for Tom Selleck."

She had little use for space flight participants, she told friends. She believed congressmen were just "invading our space," and asked, "What are we going to do with these people?" But she admired the teachers. She had even grown fond of them.

"Listen," Resnik said to Christa and Beres, "that medicine they gave you is too strong. You'll be much more comfortable later if you only take half of it."

"It's too late," they said in unison, looking at each other as if they really had poisoned themselves. Resnik shook her head consolingly and moved on to try to warn the others.

At 9:00 A.M., a half hour before the flight, Christa slipped off her pink shoes and pulled on a pair of black combat boots. Wearing a green fatigue jumpsuit over her pink and white blouse and pink cotton pants, she stood with Beres in the dressing room, rocking from foot to foot, waiting.

"Are you hot?" she asked a few minutes later.

Beres said she was, so they climbed out of their flight suits, took off their street clothes and pulled on the green fatigues again. Meanwhile, everyone else was on the jet. Seven of them had stridden onto the runway as a group, led by Dick Methia, who had torn open his jumpsuit to reveal a T-shirt that said I'M NEW BEDFORD'S OWN PETER PAN. Then Morgan had arrived, clicking her Nikon camera at the gaggle of photographers who stood by the jet's boarding ladder. Now the eight of them were wondering what had happened to Christa and Beres. So were Christa and Beres.

After their quick change, they had dashed out of the dressing room and headed in the wrong direction, sprinting down a long corridor and into a hangar for NASA's supersonic jets. A mechanic had pointed them in the right direction, but then they had remembered the instructor's orders not to run and it had taken them twice as long to reach the runway. Christa shoved a piece of Big Red Dentyne gum into her mouth and managed a smile as she hoisted herself up a steel ladder to the cabin. Behind her the jet's giant engines whistled and spewed vapors that squiggled in the Texas heat. Painted on the jet's belly was THE

WEIGHTLESS WONDER. Near the inscription was a black rose.

Within minutes the jet soared six miles above the Gulf of Mexico and turned into "the granddaddy roller coaster of all time." All but Christa, still haunted by her experience on the spatial disorientation chair, unbuckled themselves from their seats in the rear of the jet and moved into a hollow cushioned area of the cabin that stretched thirty feet toward the cockpit. As the jet dived they struggled against the increased force of gravity. Then the nose rose suddenly and they were free. They were floating.

Resnik and Dale Gardner, another veteran astronaut, emptied a bag of toys — tennis balls, paper airplanes, pieces of string, a water bottle, a Frisbee — and the teachers frolicked in forty seconds of free-form aerial gymnastics, bouncing off the walls, somersaulting through the cabin, skimming a Frisbee across the ceiling. Christa sat out several more dives until she felt comfortable enough to climb out of her seat and clutch a railing. Her air sickness bag stuck out of her pocket.

She tried touching her nose on the next plunge, but her arms were tugged down by the force of gravity.

"It was like I had just failed a sobriety test," she said.

In weightlessness, Christa's hair rose, her feet left the floor and her pendant floated above her head. Her arms became wings.

"*E-e-e-eow*," she screamed, amid a symphony of "All rights!" and "Yahoos!"

On the next few dives she tossed a tennis ball that twisted and turned so often that John McEnroe couldn't have touched it. She launched a paper plane, tossed a Frisbee and released a piece of string that rose as if it were hitched to a helium balloon. She bounced from wall to wall with the touch of a finger, winging through the cabin like a swan in slow motion, rolling in graceful loops through a traffic jam of anchorless teachers, her brown eyes wide with wonder. She held hands with the other teachers and formed a floating circle in a free-flight show of unity.

Then she felt ill. The jet dived twenty-seven times in two hours, and Michael Metcalf, the former Air Force pilot, had lost his breakfast early in the flight. Ladwig had fared no better, and now Christa felt queasy. She returned to her seat, her pride aching as badly as her stomach. Tears welled in her eyes. Beres sat beside her.

"Are you all right, Christa?"

Christa said nothing. She held Beres's hand.

217

"It's okay," Beres said. "It's almost over. We're going down now."

Christa composed herself on the descent and smiled for the cameras on the runway. Then the flight crew whisked the ten teachers away. It was lunch time, and they knew just the spot — Pe-Te's, a former gas station across from the air base that had been converted into a Cajun barbecue-and-beer restaurant.

Resnik, like most of the astronauts, was a regular at Pe-Te's. She and Ellison Onizuka, who organized monthly beer parties at the restaurant, had volunteered as judges in a gumbo cook-off the owner, Les Johnson, had held two months earlier to raise money for Houston's public radio station. The cook-off, which was held in nearby Clear Lake Park, had drawn more than eight thousand people and raised $13,000.

Autographed pictures of former and current astronauts appeared throughout Pe-Te's, but the dominant decor was a brightly colored collection of five thousand vanity license plates, representing fifty states and eighty-nine countries, that Johnson had tacked to the walls. One of the Texas plates said SHUTTLE. A large wooden chicken wearing a saddle sat in a corner for children to ride, and from the ceiling hung primitive utensils — spoons, a

218

slop jar, a butter churn, a plow — that Johnson called his junk collection. As the flight crew stepped from the sweltering concrete parking lot into the bone-tingling chill of the air-conditioned restaurant, Johnson greeted them by name. The teachers wore their flight suits. The air was thick with the smell of barbecue.

So what if I had felt a little queasy on the vomit comet, Christa thought. She had made it, and that was all she needed to know. As the others ordered barbecued crawfish, a Cajun sausage dish and Resnik's personal favorite, beans and rice, Christa rolled up her gum and decided on a beer and a plate of barbecued gumbo.

"If you told me this morning I'd be eating *anything* for lunch," she said, "I would have told you you were crazy. Now look at me. Maybe there's hope yet."

As Christa sopped up the last traces of barbecue sauce, Johnson approached the table.

"Any of you folks named Christa?" he wanted to know. She raised her hand.

"Well, you've got a phone call, ma'am."

"Who on Earth . . . ?"

It was Steve. Unable to control his curiosity, he had called the space center to ask her how she had fared on the flight. The people at the space center had told him to try the air base, and the people at the air base had told him to try Pe-Te's. Now here she was telling him the whole story, queasiness and all. He teased her for a while and congratulated her, and she promised to call him later.

After she had hung up, Christa searched for Johnson to tell him how much she had enjoyed the gumbo. She lived in New Hampshire, she said, where no restaurant served Cajun food and very few people even knew what it was. He told her that was all right, because he had never met a Texan who cooked "clam chowder soup like you Yankees can." He said he hoped to get to New Hampshire one day to find a friend he had met in the Air Force and had last seen in World War II. Christa told him to stop by for some chowder when he arrived.

"She was one fine ol' gal," he said. "And she was tickled to death about a chance to get on that shuttle."

The teachers changed into their street clothes after lunch and faced the press in an auditorium at the space center's museum. Two dozen reporters sat in the front of the auditorium, and twice as many tourists, many of them children, sat in the back. The teachers sat in alphabetical order on a stage where

America's greatest heroes had once sat, and in a manner for which reporters have become notorious, a woman from *Time* asked the first question.

"Would everyone who got sick on the KC-135 please raise their hands?"

A pale Metcalf smiled weakly.

"Okay," the reporter said, "now tell us about it."

With a teacher's patience Metcalf explained that he had never become ill as an Air Force pilot and that the KC-135 had been a new experience, not an entirely enjoyable one, but an interesting one just the same.

Then his colleagues described the fun Metcalf had missed. Bob Foerster said he had tried juggling in weightlessness. Dick Methia said no carnival ride in the world could top the KC-135. Peggy Lathlaen said the adventure had taught her something.

"I always thought the astronauts were brave and strong and had the right stuff," she said. "What surprised me was that I could do what they could do."

One by one they talked about what the ride had meant to them. When it was Christa's turn, she talked about what the teacher-in-space program could do for everyone, most of all young people.

"I can't wait to get back in the classroom," she said. "I wish my students were here right now. There's so much I want to tell them, and I can't wait to share it. Between the ten of us, you're going to have some really excited kids next fall."

As a history teacher, she said, "I want to show my students how the space program connects with them, how it belongs to them. If students don't see themselves as part of history, they don't really get involved. I want to bring them up in the space program, and maybe if they see me as part of history, it will help them share that with future generations."

When the press conference ended, the children in the back of the auditorium bounded down the aisles to the stage, eager for autographs and full of questions: "When are you going up in space?" "How long does it take to get there?" "What do you eat?" "Can kids go?"

For the first time, Christa saw the teacher-in-space program reaching children.

"Suddenly, it was working," she said later. "Kids were getting excited. They can't relate to astronauts, but every one of them can relate to a teacher, and when they saw teachers getting ready to go into space, it became very real to them. They felt like they were part of it. The program was working."

After they had signed their last autographs, the teachers hurried to Mission Control to watch the final minutes of the forty-three-hour countdown to *Challenger*'s eighth flight. Only senior NASA officials, a flight crew's immediate family and representatives of companies with cargo aboard the shuttle were normally allowed to watch a launch in the heavily guarded communications room, but NASA had made an exception for the teachers. They eased into plush theater seats at the back of the room as the countdown continued.

Before them sat thirty men and women at blinking control panels, exchanging information with the shuttle crew, the launch control staff in Florida and a ground support team of three hundred to four hundred around the world. A ten-foot-high video screen at the front of the room showed *Challenger* poised for lift-off. Computers named Scotty and Uhura — the support team on television's *Star Trek* — stood at opposite corners of the room, processing millions of bits of information a second. Several large digital clocks monitored the countdown.

The teachers had already toured the center, so they knew once *Challenger* departed a tiny symbol would track its course on a sprawling ten- by twenty-foot map of the world, moving about an inch, or 75 miles, every fourteen seconds. The ship would orbit at 17,500 miles an hour, and eight on-board video cameras would beam back pictures of its journey, depending, of course, upon its getting safely off the ground. It was a big if.

Named for a ship that had explored uncharted areas of the Atlantic and Pacific a century earlier, *Challenger* had flown twenty-one million miles in seven flights, the most in NASA's fleet. But it had encountered problems. A cracked fuel line had caused a two-month delay of its first flight in 1983. Eight months later fiery gases had come within three seconds of burning through the insulation in one of its solid rocket boosters and causing a catastrophic explosion. Beyond that, several design features once considered essential to manned space flight had been abandoned, among them a safety system that would have allowed the crew to cut loose from the giant fuel bomb during the first two minutes of a launch. The *Challenger* crew was about to be hurled toward space with 6.5 million pounds of thrust, enough power to light up the entire eastern seaboard, and they had no way to escape once they left the ground.

Christa at six years old

The Corrigan children (*left to right*): Lisa, Betsy, Steven, Christopher, Christa

McAuliffe family portrait

Christa at home

hrista teaching women's
istory at Concord High School
 spring 1985

Christa conducts the Nevers Band on Christa McAuliffe
Day in Concord

Christa describes weightlessness to members of
the Concord Rotary Club in August 1985.

Christa toasts her selection for shuttle flight with friends in her home on the night she returned from the White House.

Christa and Caroline on the night she was selected for the space ride

NASA

Christa in training: experiencing weightlessness

Christa boards KC-135 for her first experience with
weightlessness as one of the ten finalists in Houston.

Crew of the space shuttle *Challenger*. Crew members are (*left to right, front row*) Michael J. Smith, Francis R. (Dick) Scobee, Ronald E. McNair; (*left to right, back row*) Ellison S. Onizuka, Sharon Christa McAuliffe, Gregory Jarvis and Judith A. Resnik. All are astronauts except McAuliffe and Jarvis.

NASA

Bob Crippen knew that, which was one reason his heart rate reached 130 beats a minute on his first shuttle launch (his partner, John Young, a veteran of the Apollo launches, had maintained a normal heart rate). Crippen watched somberly now with the teachers as Commander Gordon Fullerton and a crew of six waited in *Challenger*'s cabin for their ride to begin. Christa leaned forward and rested her chin on her hands, glancing from the countdown clock to the team of controllers and the video screen.

At 4:30 P.M. a voice from the launch control center began the final phase of the countdown: "Ten, nine, eight, seven . . . "

With six seconds to go, *Challenger*'s three main engines ignited in rapid succession, roaring and spewing orange flames. Three seconds later a sensor detected a faulty coolant valve and suddenly the engines shut down. *Challenger* trembled on the launch pad as hoses cooled the engines with thousands of gallons of water, sending a giant steam cloud billowing from beneath the vehicle. Fullerton ordered the crew to unstrap their flight harnesses and prepare for an emergency escape.

In the Mission Control viewing room, the teachers sat mesmerized, waiting for the danger to pass. Then Crippen tried to describe the crew's frustration, and Ladwig mentioned the risks of space flight before asking if any of them wanted to withdraw from the competition. None did.

"It was more reassuring than frightening," Christa said of the aborted launch. "It told me the computers did their job. It showed me the system worked."

That night, their last in Houston, the teachers signed each other's pictures to make sure they had an autograph of the first teacher in space. Then they packed for a quick trip to the Marshall Space Flight Center in Alabama before entering the last leg of the space race — a flight to Washington for their final interviews and the announcement of the winner.

Each of them was still in the running, but none of them knew where they stood.

"Nobody ever said, 'Well, you have two hundred fifty points and Barbara Morgan has two hundred forty,'" Christa said. "And we didn't want to know. We didn't need that kind of pressure. We didn't want to play that game."

But the reporters did. The overwhelming majority predicted a woman would be chosen for her public relations value, and Christa was one of three women they mentioned the most often. The others were Barbara Morgan and Kathleen Beres.

In Concord, no one who knew Christa doubted NASA would choose her, and her husband couldn't leave his law office without someone stopping him on the sidewalk or shouting to him at a red light.

"Hey, Steve, she's gonna do it, huh?"

"Hey, Steve, she's the one."

"Hey, Steve, are you ready for this?"

He told them anything could happen, but that if Christa was chosen, he was ready, willing and able to hand over the child-rearing and housekeeping chores to his mother and his mother-in-law. He said he had enough trouble getting himself out of the house in the morning.

Now his wife was headed for Huntsville, a once-sleepy cotton town soon to be struck by a high-tech tragedy.

The idea was to give the teachers a treat after the tumult of training and before the tension of their final interviews. Still weary from the week in Houston, they arrived for a celebration in Huntsville, home of the Marshall Space Flight Center and birthplace of the booster rockets that sent Alan Shepard aloft, Neil Armstrong to the moon and a fleet of space shuttles safely into orbit. It was July 13, 1985, the space center's twenty-fifth birthday, and the teachers joined three thousand employees and townspeople for an anniversary picnic in the steamy summer heat.

Before they had digested their hot dogs, Christa and her companions were whisked across a parking lot to sample the rides at the U.S. Space Camp. Dressed in astronaut blues and space camp baseball caps, the teachers each landed a shuttle simulator on a video runway and placed an American flag on a mock moon by riding a mechanical arm to the roof of the domed camp. Christa laughed and flashed the thumbs up sign as she planted the flag, her colleagues laughing and waving to her. They were relaxed. The vomit comet was behind them, a day of rest lay ahead and now, to their delight, they were about to meet a celebrity — country singer and space fanatic John Denver.

Denver had traveled to Huntsville for the Marshall picnic. He was due at the airport for a flight to Washington and a concert with the Boston Pops, but he was so tickled to meet the teachers that he spent nearly an hour chatting with them, asking them why they had entered the space contest, how they qualified for the final ten and what awaited them. He annoyed several NASA officials by claiming to be the driving force behind the Space Flight Participant Program, but he flattered the teachers by promising to write a song about them. He gave each of them his home address in Colorado and asked them to send him their teacher-in-space applications.

"Can you believe it?" Christa said. "*John Denver* wants to know all about us. He wants to put *our* thoughts to music. Hopefully, we can give him some wonderful inspiration."

When Denver left, the teachers toured the Alabama Space and Rocket Park, where they marveled at the size of the boosters (fifteen stories high and twelve feet wide), which hurled the shuttle into space. Then they tried two more rides, neither of which appealed to Christa and her squeamish stomach. The first was a multiaxis chair that rotated more violently than the spatial disorientation chair in Houston. None of the teachers was anxious to try it, but none of them dared to pass it up. They knew NASA was watching.

"We were afraid if we sneezed the wrong way we'd be in trouble," said Kathy Beres. "We couldn't take any chances. The stakes were just too high."

All ten tried the chair and only two became ill — Bob Foerster and Christa. Foerster went to lie down in a nearby dormitory while Christa stayed, trying to recover in time for the next ride — the Lunar Odyssey, a centrifuge that spun at about 35 mph to simulate a gravitational force stronger than shuttle riders felt upon lift-off and strong enough to turn Christa's stomach again. Pale and weak, she listened to Gregory Walker, a twenty-year-old summer employee at the rocket park, explain the excitement of the ride.

The Lunar Odyssey simulated a flight to the moon, Walker said. A circular chamber would spin, pushing the occupants back in their seats as a narrator described the journey and video images of the mission flashed on the domed ceiling. The riders would experience the roar of the lift-off and the silence of space travel. The adventure would last about fifteen minutes.

Christa waited outside while her colleagues climbed aboard. She decided to sit out the ride, then changed her mind at the last minute, determined not to let her stomach conquer her spirit. She strode into the chamber, fastened her seat belt and the ride began. The teachers sat near two thin walls inside the centrifuge. Gregory Walker sat near a control station in the middle.

Gaining speed, the centrifuge soon created twice the force of gravity, a sensation the teachers had felt on the KC-135. Now they sat in the darkness, listening to the narrator as images of the launch unfolded on the screen above them. They had cleared the tower, blasted through the atmosphere and then, just as the narrator announced they were "about to escape Earth's gravity," they heard a loud, unexpected thud, followed by a rhythmic thumping noise. They heard employees screaming.

"Stop it! . . . No, don't stop it! . . . Slow it down! . . . Where is he? . . . Call an ambulance!"

Confused NASA officials took a quick head count and discovered one of the teachers was missing — Foerster.

"Oh my God!" Alan Ladwig shouted. "It's Bob!"

Then someone reminded Ladwig that Foerster had not entered the chamber. The teachers were safe, but Gregory Walker, a graduate of Huntsville's Virgil Grissom High School, was not. He had made a mistake. Instead of staying strapped in on the ride, rocket park officials said, Walker had climbed about the chamber, fallen and been hurled through one of the thin walls into the machinery below.

Now the ride was slowing to a stop, and the employees, the teachers and their NASA escorts were searching for him. There was shouting, confusion, then a shaken voice — Christa's.

"He's over here," she said.

She had discovered the young man seriously injured and unconscious, pinned in the machinery near her seat. He was taken to an ambulance while Christa, choking back tears, was rushed from the building with her colleagues. Several hours later, after the teachers had listened to Konrad Dannenberg describe the power with which the booster rockets would propel one of them toward space, they learned Walker had died. Christa carried the image of the accident with her for the next seven months.

They boarded a NASA jet just after dusk for a flight from Huntsville to Washington, where only forty-five minutes of the next five days really mattered. In alphabetical order — Beres to Wenger — the teachers would walk three blocks from the L'Enfant Plaza to NASA's white marble headquarters and face the space agency's brass in their final interviews. Each of them would get forty-five minutes, no more and no less, and Christa would make the walk on Wednesday.

She had three days to wait and only Concord would have been a better place for her to spend them than Washington. Some of her best memories were in Washington: her first years of marriage, her first teaching job, her first house, her first child. She had made the final ten there. The city had been kind to her, and as she washed her laundry in a friend's house on Sunday afternoon, she imagined it giving her one more glory day.

"I'm feeling pretty good about my chances," she said. "I haven't done anything really bizarre to pull myself out of the running, but it's hard to analyze exactly what's going to affect their decision — maybe your background, your family or what part of the country you're from. I don't know, but I feel confident."

She toured the Goddard Space Flight Center in Maryland on Monday, home of the $100 million communications satellite *Challenger* would carry into space in January. On Tuesday, she posed for pictures with her congressmen — this time they cared — and met with the members of the Air Force Association to discuss aerospace education. On Wednesday, she made the walk.

Christa met no actress or basketball star on this trip. NASA's final selection committee consisted of seven senior officials, among them the general counsel, the external relations officer, the equal opportunity supervisor, the space flight director and Ann Bradley, the committee's no-nonsense chairman. Like the earlier judges, they had studied the videotapes and the applications of each candidate, but none of that mattered much anymore. What they needed to know was which one of the finalists could best promote the space program and handle the pressure — the time away from home, the media, the training, the flight. Once and for all, they planned to find out.

"Some of our questions were intended to throw them off guard so we could see how they reacted," Bradley said. "We had never chosen someone like this before, and we knew whoever we picked was going to get a lot of attention. Some reporters aren't too kind, you know."

Neither were some of the judges. They had pressed Beres so hard that Bradley later apologized to her. Foerster, Garcia, Lathlaen and Marquart had taken the same heat, and now came Christa.

Most of the questions were familiar: "What's going to happen to your family?" "How would you feel about living out of a suitcase for several months?" "Do you see yourself as an inspiration to people?" "How do you know the shuttle won't scare you?"

Christa provided the same answers that she had provided in earlier interviews, but this one was different. The other judges had accepted her answers. These judges confronted her.

"What do you *mean* by that?" they wanted to know. "What do you mean you *think?* Don't you *know?*"

Had she not been the captain of her college debate team and a high school law teacher who understood the art of questioning a witness, Christa might have buckled under the pressure. Instead, she paused, reconsidered their questions and calmly qualified her answers. They liked her. She was articulate and enthusiastic, "the one we thought could talk to five thousand people at the National Education Association convention and impress them," Bradley said.

Not only was Christa gifted, she had also done her homework.

"All ten teachers were outstanding people," Bradley said, "but some of them concentrated too much on how wonderful it would be to fly in the space shuttle, rather than on *how* they would use the experience to get teachers excited about the space program. Christa was the one who most clearly understood what we had in mind."

It didn't matter that she was a woman from a state with the nation's first presidential primary, a potential boon for Vice President Bush, or that she came from a state whose school system thrived with only token federal aid, a source of pride for President Reagan.

"Politics never came up," said Robert Nysmith, a member of the selection committee. "It was a zero consideration."

Bradley had blocked out several hours for the judges to deliberate, but they needed only a few minutes. Their first vote was unanimous. Christa was going up in space.

CHAPTER SEVEN

IN A WORD, said the headline in the *Concord Monitor*, WOW!

The hometown girl had done it. Right there in the Roosevelt Room, Mrs. McAuliffe, the field-trip queen of Concord High School, had won the greatest field trip of all. She had shot from obscurity to national celebrity on her way into history, and nowhere but Concord had she stirred more pride.

Shopkeepers wept with shoppers as the news raced down Main Street during the city's annual Old-Fashioned Bargain Days festival. Teachers hugged and students traded high fives. Parents picked up their small children and explained that a neighbor, someone just like mommy and daddy, had achieved an American dream. A hot dog vendor shut the lid on his steam table and joined the celebration. Almost everyone claimed to know the triumphant teacher or to know someone who knew her.

EARTH TO CHRISTA, said the city's weekly newspaper, HAVE A NICE TRIP.

Across town, Matt Mead, a student in Christa's law class, rushed next door to share the news with Patricia Smigliani, a classmate who lay sunbathing in her backyard. Christa's victory was their victory, and they romped like children through the neighborhood.

"She did it!" they shouted. "Our teacher did it! *Our* teacher did it!"

Even before she had left the White House, Christa fever burned hot in Concord. Grocers unfurled WAY TO GO, CHRISTA! banners. Librarians decorated the children's room with books about Sally Ride and the space shuttle. A toy-store owner rushed her space merchandise to a display window. Travel agents began booking trips to the launch.

"Guess what!" cried an eight-year-old girl, gulping air as she sprinted toward a group of dirty-faced friends in the south end of town. "My mom's gonna take me to Florida. We're gonna go see Christa fly."

And for every smiling face, it seemed, there was a photographer. Newspapers from Hackensack to Honolulu planned to trumpet Christa's victory in their morning editions and they needed a little hometown color. The city was under a media siege.

A senior partner at Steve's law office directed traffic in the parking lot as reporters by the dozen descended upon "the happy hubby," as he was described by the *Boston Herald*. Once inside they unloaded an arsenal of questions, only two of which really seemed to intrigue them.

First, "How do you feel about losing your wife for a year?"

"I'll miss her a lot when she's gone," Steve said, "but that's a small price to pay for an experience like that. And I think it will be great for our grandchildren. They'll appreciate it more than anybody."

And "Are you worried for her safety?"

"Surely there is danger," he said, "but no reasonable human being would worry about that for a chance like this. Who wouldn't want to go?"

By nightfall Steve had answered so many questions about his wife that he "suddenly felt a strong affinity for Prince Philip." The role was familiar though, he joked, because "I've spent my entire married life known as Mr. Christa McAuliffe."

MR. CHRISTA said the headline in the *Union Leader*, the state's largest newspaper.

The media caravan rolled from the law firm up The Hill to complete its portrait of the space teacher's family. Television helicopters buzzed her home, reporters tripped over each other on the front stairs, photographers shot every possible picture from every possible angle and then shot them again. The telephone rang and rang, and Steve's mother, Rita, tried to keep track of where the calls had come from — New York, Florida, New York, California, New York, New York . . .

"It's been wild," she said, laughing, a can of Coors sweating in her hand. "It's amazing to think Christa's so famous. I just hope I'm around to see her name in the history books."

Swinging a teddy bear in one hand, a bag of popcorn in the other, Caroline greeted reporters at the door in her bathing suit.

"I saw my mommy on television," she told each of them. "I really liked it. It was neat."

When one of them followed her into the kitchen, she shoved a handful of popcorn into her mouth and introduced him to her grandmother. Then she told the reporter she was five years old and could barely wait to see her mother take off on the shuttle.

"They go way up in space," she said, pointing out the kitchen window.

Scott, meanwhile, had lost his enthusiasm for the press.

"Nana, when is the phone going to stop

ringing?" he asked just before sunset. "When are they going to stop asking me questions and taking my picture?"

"Just think," Rita told him, "when you go to school in September, you're going to be the most popular kid in Concord."

September? Who cared about September? He grabbed his baseball glove and retreated to the backyard to play catch with Roger Jobin.

"Why don't you go out and talk to them?" Jobin said as another television crew turned into the driveway.

Scott threw back the ball and shrugged. "What for?"

All that mattered was that his mother was coming home, and the sooner she got there the better. Two weeks had been a long time to be gone.

Christa knew that, of course, and she would have been home sooner had NASA not asked her to stay in Washington and dash from the banks of the Potomac to the front lawn of NASA headquarters to the roof of the Hall of States Building, from one makeshift television studio to another, to tell the nation why she wanted to pioneer space for the common man. Getting stuck on the elevator at the Hall of States Building hadn't helped.

"Are we in the Twilight Zone?" she asked

241

the others on board — NASA's Ed Campion, a television producer, a New Hampshire senator and his aides. When one of the senator's aides said, "I hear the same people who built the elevator built the shuttle," Campion cringed.

By the time she had finished her last interview, eaten a taco salad and changed planes in Hartford, Connecticut, for a flight to Manchester, New Hampshire, twenty miles south of Concord, it was nearly eleven o'clock and Christa was tired — tired of talking and tired of traveling. She was counting the minutes until she could lie in her sunken tub near a top-floor window with a view of the stars.

"It seems like I've been gone for ages," she said.

As the plane approached the runway, she wondered if it really had been only thirteen days since she rode the freeway south from Houston. She wondered if all of this really had happened to *Her,* Sharon Christa Corrigan McAuliffe, the woman who had trouble getting to school on time. She studied her face in a pocket mirror for a moment. Yes, she told herself, you really are *going up in space.* She patted her hair and refreshed her lipstick as the plane taxied toward the terminal.

A mountain breeze met Christa on the tarmac. Spotting Steve and Scott, she rushed to embrace them, still wearing her lucky yellow jacket with the red rose in the lapel, a reporter from *People* magazine hot on her heels.

"You won't believe it," Scott told her, glancing toward the terminal.

The little reception she had expected had swelled to several hundred well-wishers, among them her daughter, her parents, her brother and sister, her in-laws, a small army of reporters and photographers, a couple of politicians, a kilted bagpiper and two people dressed in pig suits.

Who cared if no one in the room knew a Cajun meal from a decompression chamber? So what if her back ached? She was back in New Hampshire, home among friends and family. She burst upon them smiling from ear to ear.

"I'm so glad to be home," she said. "I wish you all could have been there."

The bagpiper played and the pigs danced. Balloons and WELCOME HOME signs bobbed above the crowd. Her mother gave her a dozen red roses, and her father, who had retired that morning from a thirty-year accounting career, swept her into his arms. Caroline wore her jelly shoes and a half dozen plastic bracelets and clutched her mother's jacket as reporters pushed closer. Photographers hollered, "Hey, Christa, how about a kiss for Steve."

She obliged them, and then they wanted a kiss for Scott.

"Yuk," Scott said, wiping his cheek. "Too much lipstick."

Autograph hunters and interviewers kept her busy until midnight. Then, with Roger Jobin's video camera trained on her, Christa squeezed into her Volkswagen bus with family and friends to follow a police escort twenty miles up the six-lane interstate to Concord.

"Whew, am I glad that's over," she said before the motorcade had left the parking lot.

In Concord, the police had no trouble finding her house, the only one in the neighborhood with all the lights on, a CONGRATULATIONS sign sticking out of the front lawn and balloons in the trees. In the house were more balloons, flowers, telegrams, a sign that said OUT OF THIS WORLD, champagne, friends and photographers. More photographers.

What is it with these photographers? she wondered. Dick Methia must have known what he was talking about when he said, "My heart goes out to Christa and her husband and her kids because I have felt something of what they are going to feel. It's the downside of celebrity status, and it's not easy to take."

Cameras clicked from the first champagne toast until Christa slumped onto the carpeted stairs and leaned against the wall with a yawning Caroline in her lap and Scott, holding his stuffed frog, Fleegle, trying to keep his eyes open on the step behind her. It was long past midnight and everyone but the photographers, it seemed, was ready for bed. Most of them left when Christa excused herself. However, one of them insisted on lingering on the front porch until the police removed him.

Christa was a star, no question about it. Cable News Network had broadcast the entire White House ceremony to the nation. The three major networks had featured her on the nightly news and wanted her for their morning shows. Johnny Carson wanted her. *People* magazine wrapped its arms around her. ("Cut! Print!" it reported. "You can hear America thinking, 'Christa, this could be the beginning of a beautiful friendship.' ") Thousands of trees had fallen in the crush to get her story in print, with newspapers in New Hampshire the most enthusiastic of all. The *Concord Monitor* had devoted four full pages to the news. ("What a story for the community!" wrote the editor, Mike Pride. "What a day to be a journalist!") The *Union Leader* announced that Christa and New Hampshire had rendezvoused "on cloud nine." And even the Boston papers had gone bonkers.

TEACHER'S STAR TREK DREAM COMES TRUE screamed the *Boston Herald,* which ran the photographic story of her life, including her baby picture, under the headline, THROUGH THE YEARS WITH OUR TEACHERNAUT. Among the photos were a family portrait, her wedding picture and a snapshot of her at the age of two. "Cute little Christa shows off her winning smile and dazzling eyes," the caption said.

Some of the hype amused her, like the story describing her as a former all-American on the Marian High School softball team. Not only was she not an all-American, but Marian had had no softball team when she was there. Her greatest softball achievement was pitching for a successful church league team.

"I laugh when I read some of these things," she said. "Sometimes it's hard to believe it's me I'm reading about."

All the hoopla reminded people of Alan Shepard, a New Hampshire native whose mother still lived in Derry, thirty miles south of Concord. Shepard had been showered with confetti in the biggest parade in New Hampshire history when he returned from his historic flight. The state's Republican party had pleaded in vain with him to run for the U.S. Senate. His high school, Pinkerton Academy, had changed the nickname of its sports teams to the Astros. The state legislature had failed in its attempt to rename Derry "Spacetown USA," but it had named a highway for him and hung his portrait prominently in the state house. And twenty-four years later, children in Derry still turned out on the anniversary of his flight to celebrate Alan B. Shepard Kite Day.

"I'm proud New Hampshire has two historic firsts in the space program," Shepard said from his office in Houston. "We've come a long way from my little popgun flight twenty-four years ago. Christa will be the first true passenger in space. Within ten years, we'll have space stations, and within fifteen or twenty years we'll have people traveling to them on a regular basis. Christa will have set the pace."

On his way to the moon in 1971, Shepard had seen rainbows from the heavens, watched lightning dance on the dark side of the Earth and coasted a half million miles in weightlessness. He had seen "the wonders of the universe," but he had no intention of spoiling the picture for Christa. He wanted her to fly with a clean canvas.

"All I want to tell her is that it's going to be a lot more fun than she ever dreamed it would be," he said. "She's going to love seeing the beauty of the Earth from that point of view."

Nothing about space flight, particularly his first flight, had irritated Shepard more than the training, the unrelenting physical and psychological pressure, the tedium of flying 120 simulated missions and preparing for every eventuality at every stage of the journey.

"There was too much crepe hanging and hand wringing in those days," he said. "Too many people were making bad predictions about how we would respond in space. They were concerned about things to an unjustifiable degree. We don't see that anymore."

Christa would have no fear of flying, he said. "If there was that much of a trauma, she would not be the right type of person."

She would not have the right stuff.

Shepard, sixty-two years old and approaching retirement, was a wealthy businessman with a wife, two daughters and six grandchildren. But as he imagined Christa opening a new era of space flight, he itched to be part of it.

"Maybe when I've grown a long gray beard and I'm hobbling around on a cane," he said, "they'll feel sorry for me and let me go up one more time for old times' sake."

Only the media pressure would change his mind. "It gets kind of overwhelming," he said, "as I'm sure Christa's finding out."

Sure enough, before the birds had waked her the next morning, reporters and photographers had gathered in the small park in front of her home. This time, however, the police held them at bay, advising them that Christa and her family would hold a press conference at the Ramada Inn before she climbed into her astronaut blues and served as the honorary chairman of the city's annual Lions Club parade.

Yes, Christa thought, it is kind of overwhelming, this great insatiable beast that stalks me morning and night. It was nothing like she had described to her students when she was the faculty adviser of *The Roadrunner*, the school paper at Thomas Johnson Junior High. But she knew her mission would fail without the press. If she was to excite the nation about education and space, she needed to excite reporters first. She quickly adopted them as her new students, and her press conferences became her classroom.

"I'm answering the same questions over and over again," she said at the Ramada Inn, "but that's what teachers do. Every year we get a new set of kids excited about Christopher Columbus. It might be the fourteenth straight year or the twenty-fifth year, but that's our job. That's why we're there."

So here she was talking again about John Kennedy's message that everyone had the power to make a difference. Here she was promising again to return to teaching, despite her superintendent's belief that opportunity would lead her elsewhere. Here she was joking again that next year's course on the American woman would start with her mission and work backward.

And here was Steve again.

"What's this about cornflakes?" the reporters wanted to know.

"Actually, the whole story is that I come home from work very late quite often, and Christa has a standing rule that after eight-thirty, I'm on my own," he said. "Being basically lazy, cornflakes are the easiest, most filling thing."

How strange this all was, how absolutely crazy. Sixteen years to the day after they had cheered the first moon landing from a rain-slickened highway in Pennsylvania, here was Steve talking to the nation about cornflakes and Christa getting ready to climb onto the back of a Mercedes-Benz convertible for a triumphant ride down Main Street.

"C'mon kids," she said, and Scott and Caroline climbed on with her.

It was a Norman Rockwell kind of day. The sky was baby blue, and the state house dome shimmered in the sun. Flags and bunting draped buildings along the route. Babies in bonnets sucked their fingers, balloons dancing above them. Older kids ate cotton candy on the curbs. Men and women clapped in rhythm as the band played "You're a Grand Old Flag."

"Here she comes!" a woman shouted, and everyone leaned forward.

"Atta girl, Christa!" they said. "Go for it! Reach for the stars!"

She felt awkward at first. Only yesterday, it seemed, she was just another one of the city's 350 teachers. Now she wore an astronaut's uniform and sat atop a twinkling Mercedes, riding behind three pickup trucks filled with photographers hanging over each other to snap her picture. She knew the fable about the fox and the lion. She knew familiarity could breed contempt. She knew humility.

"I didn't get where I am because I worked a lot of years to achieve it or because I was the best teacher who applied," she had said at the

White House the day before. "I had a little luck and probably a judge pulling for me here or there, so I'll never see myself as *the* teacher or *the* perfect citizen, because I'm not."

Still, she wondered what her neighbors thought now, and they quickly put her at ease. They cheered and offered her flowers. They snapped her picture and asked for her autograph. They tipped their hats and shook her hand. One of them, an elderly woman, stepped off the curb and hobbled toward her with her arms outstretched.

"God bless you, Christa," she said. "God bless you."

Christa's victory was their victory. A woman who shopped and worshiped and socialized at the same places they did, a woman who was as ordinary as they were, someone who taught their children to reach as high as you could and then reach a little higher, had done it, right there in the Roosevelt Room. At one point Christa was so mobbed by spectators that the police stopped the parade.

"Does this seem possible?" she said, throwing back her head and laughing. "It's gotta be a dream."

She cradled a barefoot Caroline in one arm, Scott in the other and the three of them flashed the thumbs up sign until Caroline lost interest and Scott's arm tired. Smiling proudly, Christa kept her thumb high, her long curly hair cascading onto her flight suit from under a red, white and blue cap with a drawing of the shuttle on it. Occasionally she giggled so loudly that people could hear her above the band.

As Christa passed the state house, a plane flew over from Laconia, thirty miles north, pulling a banner that said, GOOD LUCK, SHARON. Her mother, seated in the front of the Mercedes, looked up and smiled. So did her father, who watched from the sidewalk, and her husband, who snapped pictures from one of the press trucks. Not everyone knew what to call her, but they knew what NASA had chosen her to do. Christa had touched her neighbors' souls.

"The woman is talented, she's local and she's real," said Nancy Martell Stevenson, a former special education teacher who had left the profession because of low wages. "That's what's exciting."

But it was also exhausting. The next day, after two weeks on the road and less than forty-eight hours at home, Christa was scheduled to fly out of Manchester for that morning's network news shows in New York. Searching for a few hours of peace, she retreated with Steve and the kids to the Concord Country Club, figuring the pool was as good a place as any to find it.

She lay in a lawn chair, the palest person at the pool, Steve reading the Sunday papers next to her, the kids in the water. She was talking with friends when Caroline burst out of the pool and sprinted toward her, leaving a trail of tiny liquid footprints.

"Mommy, Mommy," she squealed with a five-year-old's urgency. "Can Barbara come over for lunch?"

"No, honey, I don't think today would be a good day."

"But I want her to," Caroline said, twisting up her face and wiggling her legs.

"Another time, honey. Not today."

"But . . . I . . . want . . . her . . . to!"

"Caroline, I've got to go to the airport," Christa said softly. "If you don't want to come to the airport, maybe you can play at Barbara's house. How about that?"

Caroline's tiny shoulders slumped in relief. "That's good," she said.

One minor crisis had been averted, but others loomed. Christa knew that. She knew Caroline was too young to understand why her mother had to fly to New York, never mind someplace in space. She knew she would be gone when Caroline read her first sentence, when she lost her teeth, when she woke up crying at night. Nothing about the space sabbatical bothered her more, but Christa was a working mother about to enter one of the most exhilarating and enriching jobs of all. She had explained that to Scott and Caroline.

"Of course, it's one thing to talk about it," she admitted, "and another thing to disappear."

She was convinced, though, that they would learn more from seeing her reach for the stars than from seeing her pass up such a chance. So at the moment she was more concerned with finding a child-care provider who was flexible enough to tolerate Steve's occasional late nights at work. On weekends, Steve explained, "I'll just have to get two extra seats for the golf cart."

Caroline slouched back a few minutes later.

"Mommy, my tummy hurts," she said.

"Oh, and you were swimming so well," Christa said, pulling her close. "How about something to drink? Would you like some toast and Pepsi?"

Toast and Pepsi? Caroline's chin jumped off her chest. She would *love* some toast and Pepsi. Christa bought sodas for Caroline and Scott, beers for herself and Steve, and a bacon, lettuce and tomato sandwich for lunch. She shared the sandwich with Caroline while Steve and Scott played catch behind them.

Between bites she talked about her journal. Christa had started it in the early days of the selection process, but a week had passed since she last opened it and the entries had grown more and more sparse. Her only hope of re-tracing her steps, she said, was to try to re-fresh her memory with newspaper clippings. She said the journal was important not only for her children and her students, but also for the book she planned to write.

"The title?" she said. "Well, if I had to pick one today, it would be *A Dream Come True.*"

A few minutes later, she had to leave. She rushed home, picked up three more outfits her parents had bought her and followed a police escort to the Manchester airport and a flight to Manhattan. Hopping from one limousine to another, Christa went on a ninety-minute media binge in the morning, speaking to the nation from the sets of ABC's *Good Morning, America,* the *CBS Morning News* and NBC's *Today* show. She told Joan Lunden on *Good Morning, America* that she "would love to go into the classroom right now and say, 'Hey kids, guess what I've been doing the last two weeks.'" She mentioned to

Phyllis George of CBS that her women's history course just might start next year with shuttle mission 51-L, and Christa smiled when Bryant Gumbel of *Today* asked her if the space shuttle frightened her.

"Maybe just a little?" he said.

"Not yet," she said. "Maybe when I'm strapped in and those rockets are going off underneath me I will be, but space flight today really seems safe."

Christa stopped in a Manhattan drugstore on her way out of town, and a woman in the check-out line studied her face for a while, puzzled. Then the woman asked, "Didn't I just see you on television?"

"Yes," Christa said, suddenly reminded that beyond the television studio walls had sat the accountant in Portland, the truck driver in Orlando, the grandmother in Queens, millions from Caribou, Maine, to Baja, California.

At that moment, said Ed Campion, who stood in the same line, "Christa started to realize just how well known her face was going to become."

Steve Ballard, Christa's substitute mail carrier, had left the post office that morning with a bag that was three times heavier than usual and grew heavier as people along his route asked him to deliver cards and letters to her home on Park Ridge Drive. He accepted them gladly, figuring the extra weight would be worth it if he got to meet her.

"Then I realized she was in New York to be on television," he said, his voice dropping an octave.

Still, he was relieved to empty his bag into her mail slot, creating a small, multicolored hill of correspondence on the other side of the door. By the time Christa, Steve and the kids returned from a week at a friend's cottage on Cape Cod, the hill had grown so steep that NASA provided a Dictaphone and a team of secretaries to help Christa conquer it. The woman who had sought serenity in Concord seven years earlier had suddenly lost it.

Some of the mail was a welcome surprise. Her fifth grade teacher, Mrs. Stapleton, wrote to her. So did Sister Hogan, her high school teacher, and Professor Haglund, who had taught her the historical value of keeping journals. Students she had forgotten years ago wrote to her. Kids and teachers from Concord High sent notes by the dozen. She heard from McAuliffes she had never met, some of whom provided new chapters in Steve's family history. Several people wrote to say they liked her idea of keeping a journal so much that they had begun keeping their own. Others wanted to tell her she had inspired them to try things they once had feared — going back to college, applying for a more challenging job, testing the limits of their potential.

They wrote from around the world, many of them simply to "Christa McAuliffe, Concord." An autograph collector who claimed to have the signatures of every astronaut from Yuri Gagarin to Gordon Fullerton wanted Christa's for his collection. Another man wanted her to sign six one-dollar bills, which she agreed to do after worrying that she would be breaking the law. The International Zucchini Festival asked her to take its prize squash aboard the shuttle. An eighty-two-year-old woman wrote to say she had been a pilot for forty years and "would dearly love to go up on that old space machine." Songwriters from New York, California and Alabama sent her their music.

One of the songs was titled "Columbia" — the wrong shuttle — but that was all right. Caroline liked it so much she learned the words and sang along:

"In my spaceship *Columbia*, I'm flying
    high and free,
I zoom across the heavens and can't believe
    it's me.

I'm on my way to somewhere where eagles
    dare not go,
To find a new tomorrow beyond the old
    rainbow."

And for every kind gesture, it seemed, came a financial opportunity. Corning Glass sent a free mug and information on the Corning Space Shuttle Sweepstakes. A sweat-shirt manufacturer solicited her. Lecture offers poured in from organizations across the country. National magazines and book publishers jockeyed for the rights to her story. People asked to write her biography. Others asked to be her agent. Film producers wanted to do the story of her life.

"Can you imagine?" she said. "What would it be? Girl meets boy, girl and boy fall in love, girl goes up in space? I'm not sure it would do very well in the ratings."

NASA, of course, had anticipated the enormous marketing potential of the first private citizen in space and had forbade her from profiting from the experience for at least a year. After that, said Administrator James Beggs, "if there are profit opportunities . . . good luck, and God bless her."

Christa knew that dozens of endorsement offers awaited her, and she planned to be selective.

"A lot of people want me to help them sell balloons and T-shirts and that kind of thing," she said, "and, well, I'm not sure about that. But I would love to lend my name to something that has to do with education — an educational fair, a readathon, something that will get students interested in learning."

Others wanted Christa to get interested in politics. First, Peter Spaulding, the executive councilor for her district, recommended Christa for a seat on the state board of education, even though the vacancy was for someone other than a teacher. Then her husband blew things wide open by mentioning the couple's loyalty to the Democratic party — "How can you be a Republican and have a conscience?" he told the *Boston Globe* — and by suggesting that Christa would be a good candidate in a state sorely lacking good Democratic candidates. Even the *Union Leader,* the right-wing torchbearer for a state in which only two Democrats — Franklin Roosevelt and Lyndon Johnson — had won a presidential election in fifty years, the newspaper in which a front-page headline called Christa's idol, John F. Kennedy, THE NO. 1 LIAR IN THE UNITED STATES on the day of his inauguration, the paper that drew a joke from Steve McAuliffe each morning when he picked it up on the front desk at his law office, even the *Union Leader* leaped on the bandwagon, describing Christa as "poised, articulate, bright, funny, down to earth and affectionate." Finally, George Bruno, the head of the state's Democratic party, started to write her notes.

"Christa is an example of the type of person we need in public life," Bruno told reporters.

Publicly, she denied any political ambition. "I don't see this as a stepping-stone to something else," she said. "When I go into a radio or TV studio, I'm looking at it from the perspective of someone who has never been in a radio or TV studio before. I'm eager to go back and tell the kids what it's like. Maybe if I had political aspirations before this, I might look at it another way. Maybe I would want to seize the moment, but right now I want to return to my classroom."

Privately Christa told friends the idea of one day holding a position in which she shaped public policy appealed to her. None of her options had much value, however, until she rode *Challenger*, and at the moment — July 29, 1985 — *Challenger* was a wounded bird.

Seventeen days after Christa had watched it stutter on the launch pad from her seat in Mission Control, the 110-ton spacecraft cleared the tower at the Kennedy Space Center and rode a seven-hundred-foot trail of fire toward the heavens, soaring from zero to the speed of sound in sixty seconds. A minute later, its giant booster rockets burned out on schedule and began their thirty-mile plunge back to the Atlantic. Hurtling at nearly 3,000 miles an hour, *Challenger* continued to gain speed, its three main engines fueled by a volatile mixture of liquid oxygen and liquid hydrogen that propelled the orbiter about eight feet per gallon. Then five minutes and forty-five seconds into the flight — three minutes before entering space — Commander Fullerton told Mission Control in a voice void of emotion, "We show a center engine failure."

A sensor had alerted an on-board computer that an engine had heated to a dangerous 1,950 degrees Fahrenheit. For the first time in the history of the U.S. manned space program, an engine had shut down in flight.

"Roger," said Mission Control. "We copy."

Rising at 8,800 miles an hour, seventy miles above the Earth, *Challenger* was too high to ensure a safe landing on emergency runways in Spain or Africa. There was a brief silence.

Then Mission Control: "Abort ATO, abort ATO" — abort to orbit.

By firing the remaining two engines 86 sec-

263

onds longer than usual, the craft reached a stunted but safe orbit 197 miles above the Earth, 50 miles lower than planned. The crew had survived the most dangerous moment of the shuttle program to date.

Christa learned of their brush with death on the television news the day after she returned from Cape Cod. As she scurried about the kitchen looking for Caroline's sneakers in a futile attempt to get her to camp on time, Christa said she was more relieved than frightened by *Challenger*'s dilemma.

"Sure, they went to a lower orbit, but the point is that *they were able to do that*," she said. "Just think of those early astronauts in those capsules who had no buttons to push. They had no control over their destiny if something went wrong. Today's program is a very safe program, so I'm not nervous. I'd just like them to get all the kinks out of it before I get in it."

Roger Boisjoly, an engineer for Morton Thiokol, Inc., the company that manufactured the shuttle's booster rockets, wrote a memo later that day to senior management warning of problems in the O rings, the rubber seals between the rocket's sausagelike joints. He worried that fiery gases would burn through the joints during a future shuttle launch.

"It is my honest and very real fear that if we do not take immediate action to dedicate a team to solve the problem . . . then we stand in jeopardy of losing a flight, along with all the launch pad facilities," he wrote. "The result would be a catastrophe of the highest order — loss of human life."

Christa knew nothing about the memo. She knew only that *Challenger* was safely in orbit and that in less than twenty-four hours she faced her most rigorous test of the summer — a visit with Johnny Carson. She had forgotten the last time she watched the show, but she remembered Johnny's devilish humor, and she knew a space-bound teacher would be a perfect target. Was she worried?

"Oh, no," she said. "If I can handle teenagers in school for fifteen years, I can survive Johnny Carson for fifteen minutes."

So in the time it took *Challenger* to circle the Earth three times, Christa flew from Boston to Los Angeles to prove it. She straightened her dress and patted her hair as she waited behind the curtain for Johnny to introduce her.

"I'm sure you have read about or heard about my first guest tonight," he said. "She has a three-page article in this week's *People* magazine. She's a high school teacher from Concord, New Hampshire" — scattered applause — "and she was recently selected from a field of over eleven thousand applicants to be the first citizen in space."

After the audience watched a tape of Christa choking back tears at the White House, she strode confidently on stage, kissed Johnny and Ed McMahon, and settled into a seat where grown men and women had been reduced to giggling, stuttering shells of themselves.

"It's too bad, Christa, you didn't get emotional during that announcement," Johnny deadpanned. "You stayed cool and collected like a real astronaut."

"Well, you know, Johnny," she said without skipping a beat, "I was so afraid the only thing I was going to get out was 'one body,' and that it was going to end at that. I thought I'd never finish."

She giggled her infectious giggle. She was poised and confident, full of energy. She was the girl next door, and more. She talked about her camaraderie with the other finalists, her courses at Concord High and her instant celebrity. She explained the teacher-in-space program, the application process and the physical requirements for space flight. Then she confronted the inevitable.

"Are you in any way frightened?" Johnny wanted to know.

"Um, I really haven't thought of it in those terms," she said, "because I see the shuttle program as a very safe program, but . . . "

Johnny raised his pencil with the eraser at both ends. He cut her off. He had a joke.

"Who was it who once said — Deke Slayton, I think — I may be giving credit to the wrong astronaut — but they asked him how he felt up there in that capsule, and he said, 'It's a strange feeling to realize that every part on this capsule was made by the lowest bidder.'"

Ed laughed. Christa laughed. The studio audience and the millions watching at home in bed laughed. Johnny tried another one.

"You know I remember a few teachers when I was a kid that I would have lo-o-o-o-ved to have seen go into outer space."

Christa fixed him with the kind of playful, disapproving smile she gave students who were one wisecrack away from the principal's office. Then she decided to play along. She told him about the claustrophobia test, explaining that one day, in case of an emergency, "people would zip themselves into these little balls and they would maintain life" — her chin quivered as she tried to keep from laughing — "for about two hours while . . . "

"While what?" Johnny asked.

"While something is happening with the craft. A rescue craft could come along or . . . "

"In other words, you're just a big Baggie in space?"

"Or, I kind of had this wonderful mind picture of the person wearing the space suit kind of carrying these two balls with people floating in them," she said, giggling, the audience giggling along.

Before her fifteen minutes had ended, Christa described the speed of the space shuttle by explaining that she could step onto *Challenger*'s exercise machine as it passed over California and jog across the United States in nine minutes.

Johnny liked her. She was calm and casual and clearly excited about the opportunity that lay before her.

"I think NASA made a very good choice," he said, "because I think you can communicate this to most of us who really can't understand all of it. You're really excited to go, huh?"

"Oh, I really am," she said. "I can't wait."

"Well, I wish you well," he said, "and I think the whole country wishes you well."

As the band played "Off we go, into the wild blue yonder," Christa shook hands with Johnny and Ed and headed backstage to wipe the makeup off her cheeks, her nose, her chin, her forehead, her neck, "even my ears," she said with pained expression.

Johnny still had a few seconds before a commercial.

"Now, *there's* an experience in life that very few people will ever have the chance to participate in," he said to Ed. "Would you like to?"

"I would love to do it," said Ed, a former marine whose weight had ballooned through the years.

"O-o-o-o-h," Johnny said, "would they need a booster for you!"

"See if I zip up your bubble," Ed retorted.

As Christa returned to her hotel, guitarist Stanley Jordan played a jazz rendition of "Moon River" and actress Rita Wilson endured a nervous stint in the hot seat. Then it was time for Johnny to say good night.

"Thanks, Rita," he said, "and thanks to Stanley Jordan, and thanks to . . . who was out here first? . . . Oh, yuh, the lady astronaut . . . thanks, Christy McAuliffe."

After she had visited her sister Betsy, a waitress in nearby Venice, California, and spoken with her brother Steve, a lawyer in Salinas, Christa flew home to complete the strangest summer vacation of her life. So much had already happened, and so much had yet to be done. Her training began in thirty days and she still needed to design her space lessons, prepare her replacement at Concord High, conduct hundreds of interviews, answer thousands of letters and spend some time with her family.

First, however, the city wanted her again. Three weeks had passed since Christa's impromptu parade appearance, and people were prouder of her than ever. They had made a run on NEW HAMPSHIRE IS PROUD OF CHRISTA MCAULIFFE bumper stickers. They had rushed to the streets to watch *Challenger* pass over the city at night. They had driven past her house so often that the police had started calling her neighborhood Beverly Hills. They had waved to her as she shuttled Caroline to day camp, Scott to Little League, Steve's clothes to the dry cleaners. An eighty-year-old woman had spoken for many of them when she called Christa on the city's radio talk show.

"I have four children who graduated from Concord High School," the woman said on WKXL. "I have seven grandchildren who graduated from there and one great-grandchild on his way. We're all just so proud of you. Our prayers and all our hopes will be with you all the way."

"Oh, thanks," Christa said. "I'll need them."

"Oh, yes," the woman said. "God love you, Christa."

The city manager sensed the outpouring of affection and had tried since the White House ceremony to honor Christa with something more significant than an unscheduled parade appearance. Finally, on August 6, Concord celebrated Christa McAuliffe Day. Volunteers were placing a double row of American flags in her honor along Main Street before Christa had finished her cup of black coffee for breakfast that morning.

"I've hung out a lot of flags over the years," said Arthur Aznive, one of the volunteers, "but this is definitely exciting. A schoolteacher's going to fly in space, and she's from our town. It's great!"

The city was to salute her at dusk during a concert by the Nevers' Band — a volunteer orchestra that dates back to the Civil War — on the state house plaza. As she spent the day making public appearances across the state, band leader Paul Giles replaced a program of Navy songs with Air Force music, and Red Whitcomb, his trumpeter of forty-one years, prepared to pull on his blue and gold uniform despite a mouthful of stitches from oral surgery.

"No way I would miss this," Whitcomb said, his jaw clenched in pain.

The first spectators unfolded their metal lawn chairs near a statue of Daniel Webster more than an hour and a half before the band struck its first oom-pah-pah. Hundreds followed, and by dusk some of them were standing in the trees or leaning on Daniel Webster. When Christa, Steve and the kids appeared with two police officers, dozens rushed toward her for hugs, handshakes and autographs.

The band played "The Battle Hymn of the Republic," and Christa tried to accommodate everyone — little girls, elderly men, a teenager wearing a single silver glove, an elderly woman wearing a Dunkin' Donuts uniform, the two police officers and a teacher from South Dakota who said she was thrilled because "I'll never get another chance to meet you." People who had forgotten to bring paper thrust checkbooks, cash register tape, even a pocketbook before her to sign. A few rushed her with cameras.

"Hey, Christa, c'mon," said her school principal, Charles Foley, as he jostled for position with a pocket camera. "Give me a chance to get this picture, would you?"

Christa was out of earshot, so she turned away as Foley wasted a picture.

272

"Son of a gun," he said, advancing the film and trying to keep his balance amid a crush of children. "Thanks, Christa."

She heard him then, and Foley snapped his picture when she turned around.

"Oh, what a fight," he said with a sigh, backpedaling out of the crowd.

Then came C. David Coeyman, the quintessential small-city mayor, a mustachioed man who smoked big cigars, wore three-piece suits with a gold pocket watch and loved to talk. He had once pulled another city's mayor down Main Street in a rickshaw after losing a bet. Another losing bet had cost Coeyman a ride through the other mayor's city on an elephant.

"It's time, madam," Coeyman said, offering Christa his arm and leading her to the stage, a converted recreation department truck. Flags snapped, the crowd cheered and Christa smiled proudly. Scott and Caroline knelt by Steve, who sat in his suit on the pavement beneath the stage, filming the ceremony with a camera he had received that week as a gift from an old college friend. He had already begun a video library of Christa's television appearances with a cassette recorder at home.

Now here came Coeyman to the microphone, flanked by Christa and fifteen dignitaries with frozen smiles.

"Johnny Carson has had you!" Coeyman bellowed. "NASA has had you and Ronald Reagan has had you! Now your hometown has you!"

Then he unfurled a proclamation and, sounding every bit like the Wizard of Oz, said, "Now, therefore do we proclaim that the city of Concord wishes to extend our heartiest congratulations and highest commendation for her perseverance in seeking and achieving the honor and distinction of being the first teacher astronaut . . . "

He gave Christa a pewter plate — the city had no key — and a miniature city flag to take into space. Then he kissed her on the cheek, throwing out his hands and roaring, "Whoa! I hugged an astronaut!"

A school board member and a governor's councilor had their say before Christa had hers.

"I thought it was an emotional experience when I was picked," she said, her voice trembling, "but that was nothing compared to this night. It's beautiful. . . . It's wonderful."

When the applause ebbed, Christa surprised everyone by uttering only three more sentences, none of them about the glories of the city, the state or the space program. She had only one thing on her mind.

"I'm delighted to be a representative of the teaching profession," she said, "but it wouldn't mean anything unless the adults out there recognized that education was important and supported their schools, unless the teachers out there truly believed in what they were doing and unless you kids out there do the best you can and get the best education you can. That's what it's all about. So when I'm up in that shuttle and I'm not teaching at Concord High School, I want everybody working real hard to make education what it should be in this country. Thank you very much."

Christa pivoted to leave the stage but Coeyman had a surprise of his own. He pushed a baton into her hand and announced that she would lead the band in "Stars and Stripes Forever." Flustered only slightly, she smiled and snapped the baton to strike up Red Whitcomb's gold trumpet and the rest of the band. The crowd stood and clapped in rhythm, Christa smiled and Caroline looked up at her, bouncing from one bare foot to another.

Afterward, Steve bought Popsicles for himself and the kids as Christa lingered in the darkness for forty minutes, signing more autographs, hugging old friends and making new ones, staying until there was no one left to meet.

"Okay, now, I think it's time to scoot," she said, gathering the family and falling in position between the police officers. She took a few steps and looked over her shoulder at the lighted plaza.

"Thank you," she said to no one in particular. "Thanks a lot."

The next day the *Concord Monitor* carried a front-page picture of Christa conducting the band. The headline beneath it said, CITY OR-CHESTRATES DAY OF DEVOTION.

All this adulation, she thought, and she had yet to do anything. She had yet to even sign a NASA contract. In fact, there had been trouble from the start. The space agency had spent months searching for its first citizen astronaut, but it had spent little time planning what to do with her once it had found her. The day Christa was selected, she fell under the jurisdiction of NASA's educational affairs division, a tiny department that had played no role in selecting her and "was overwhelmed by the whole teacher-in-space project," said Ann Bradley.

Worse, Bradley said, the people who ran the department, none of whom held much power in the NASA bureaucracy, resented the idea of plucking someone out of a classroom — someone with no math or science background — and allowing her to achieve more in one mission than its own staff had achieved in the agency's lifetime. Sending a teacher into space "kind of destroyed their little world," she said. "I don't think they were too excited about making it work."

The first hint of trouble was a letter from William Nixon, Christa's new boss, to the Concord school district four days after her selection. Nixon asked the school district to pay Christa's salary and benefits and accept reimbursement from NASA during her year in the space program. The district opposed the request because it would make taxpayers liable for her health, disability or death benefits. Lawyers from both sides haggled until the school district agreed to accept reimbursement on the condition that NASA provide a written statement accepting responsibility for her health and welfare.

Then Christa went to Washington to visit the education people and design her space lessons. Or so she thought. Instead, she found herself negotiating a contract, treated as if she were a bargaining agent for a private contractor, discussing everything from relocation and travel expenses to her teaching methods and the rights to her journal.

"It wasn't fair," Bradley said, "especially because she hadn't expected to be put in such a position."

Suddenly uneasy, Christa hired Leo Lind, one of Steve's law partners, to complete the negotiations. Christa said later that the experience had been difficult, but that NASA had learned from its mistakes, which would make life easier for the next private citizen in space. Bradley agreed.

"It wasn't a complete disaster," she said, "but we certainly suffered some growing pains."

Christa did achieve some painless progress in her visit. Working with the nine other finalists and NASA's education staff, she developed a basic outline for the two lessons she planned to teach from space: "The Ultimate Field Trip" and "Where We've Been, Where We're Going and Why." Christa also met Dick Scobee, the forty-six-year-old commander of mission 51-L.

Scobee had worried her. She knew her meteoric flash across the media sky had bred contempt among a few people in the space agency. She knew her promise to "humanize" space travel had not sat well with astronauts who also claimed to be human. And she knew she was an outsider who had been thrust into one of the world's most exclusive clubs without a vote by its members. Little had frightened her more than the thought of the crew rejecting her as a public relations ornament, a useless commodity who had yanked a seat out from under one of their brethren. She wondered just how deeply they resented her. She wondered most about Scobee.

At first glance, she thought, he seemed like the astronauts of her youth, wholesome and handsome, tall, blue-eyed and ruggedly built with a square jaw and an air of self-confidence. And right from the start Scobee left no doubt that he was in charge, that Christa was a member of a team that had been chosen for a space mission, not a joy ride.

"Those are no firecrackers they'll be lighting under our tails," he told her. "Those things are for real."

But soon he eased up, and Christa realized he was not the macho jet jockey she had feared. He was much like her, an ordinary person who had accomplished the extraordinary, the first enlisted man to rise through the ranks to the astronaut corps.

Scobee, the son of a railroad engineer, had been more interested in planes than trains as a child, and he had enlisted in the Air Force right out of high school, serving as a propeller mechanic in San Antonio. He had met June Kent there at a Baptist church hayride, and they had been married a few months later. He had been twenty years old, she sixteen.

Attending mostly night classes, he had spent six years earning credits for a degree in aerospace engineering from the University of Arizona. Then his career began to take off. He earned his Air Force wings, flew a combat tour in Vietnam and qualified for the elite Air Force Aerospace Research Pilot School at Edwards Air Force Base, the original home of the right stuff. Still, Scobee figured he was too tall (six foot one), too old (thirty-two) and too inexperienced (five years of flying) to become an astronaut. He was wrong.

When he returned from his first shuttle mission in 1984, he told the students at his former high school, "If I can do it, anybody can."

Scobee had come to NASA headquarters to review the preliminary plans for Christa's shuttle lessons. He talked with her for a while and began to see a little of himself in her, a person of modest background and modest talents who had maintained her humility in the face of extraordinary success. He liked her.

He explained that his wife was an education professor at the University of Houston and was thrilled about the teacher-in-space program, even though teachers beyond the high school level had not been allowed to apply. He told her he liked her lesson plans, and he asked her about her family. When she told him Caroline was entering kindergarten in the fall, Scobee told her she could report a few days late for training so she could see her children off to their first day at school. He admitted he might have mellowed a bit since he recently became a grandfather.

Later, he pulled her aside. "You know, shuttle missions are taken for granted these days," he said, "but this one is unique. No matter what happens, this mission will always be remembered as the teacher-in-space mission, and you should be proud of that. *We're* all proud of it."

Buoyed by Scobee's endorsement, Christa returned to Concord more excited than ever about entering training. She only wished she was as excited about entering her house. More than a month after her selection, it remained a hodgepodge of balloons, letters, newspaper clippings, NASA drinking glasses, space camp caps, model shuttles, a bumper sticker that said JUST VISITING THIS PLANET, a collection of paraphernalia that continued to multiply. Christa might have had time to clean it had the crush of reporters not been so unrelenting.

Another one arrived before breakfast on a rainy morning in August. Be patient, she told him. Steve couldn't find his raincoat, Scott wanted to invite a friend to play and Caroline, the most vocal of all, wanted her breakfast — granola cereal with blueberries. After she had obliged them all, Christa consulted her daily list.

"Okay," she said to the stranger in her kitchen, "you're from *Instructor* magazine, right?"

Wrong.

"United Press International."

A few days later it was an Associated Press reporter. Be patient, she told him. She was on the phone with a Canadian radio station. A television crew was due in a half hour. Steve was home for lunch. Caroline was fighting with two playmates at the babysitter's house. Scott wanted to go with a friend to the duck pond.

"Just make sure you don't fall in," Christa warned him, interrupting the radio interview.

Then, "Take number seventy-eight," she

said, lifting the telephone back to her ear.

She finished the interview twenty minutes later and rushed into the living room.

"Okay, what are we doing now?" she asked Ed Campion, her NASA media coordinator.

"Three things at once, as usual," he said. The television crew was ten minutes away, the reporter and a photographer sat on the couch and Christa needed to take a shower. She asked everyone to be patient.

Halfway through the television interview, Scott and his friend returned from the pond — dry — and made enough noise in the kitchen to interrupt her in midsentence. She excused herself, helped them get a drink of water and then picked up where she had left off.

"Take number seventy-nine," she said.

Meanwhile, the boys ran upstairs to watch a videotape of the space adventure *The Empire Strikes Back*. Its theme music filtered downstairs as Christa explained how she wanted to inspire children to reach for the stars. When she was done, the cameraman wanted some footage of Scott in front of the television.

"For one minute," Scott said, exasperated, "and I'm not going to say anything."

Then they left.

"What are we doing now?" Christa asked Campion, who nodded toward the AP reporter on her couch.

Christa smiled and sat down. "You know, this could really get crazy in January," she said.

Although she never buckled under the supersonic pace, Christa was ready by late August for the family's last vacation — a week at a friend's beach-front home in Old Lyme, Connecticut. During the day they swam and dug for clams and walked along the water's edge, the waves rushing over their feet. At night they smelled the sea and heard the horns of distant ships. They were together and they were happy, but even there Christa paid the price of stardom. One day she helicoptered back to Concord to speak to the Rotary Club. Another day she rode a stretch limousine ninety miles to Manhattan to appear on the Emmy Awards show. She sat with Robert Kennedy's son Joseph and England's ambassador to the United Nations, and presented an award for the best single television news story of the year to ABC's Ted Koppel.

Cable News Network later in the week named Christa one of the top three heroes of the year. The others were Lee Iacocca and the stewardess credited with saving lives during the TWA hijacking and hostage crisis. Christa was confused.

"A hero?" she said with a pained expression.

Well, how would she define a hero?

"Someone who has defied the odds by breaking a stereotype and enduring the challenge of being the first at something," she said. "Look at Margaret Thatcher. Whether or not I agree with her politics, the point is that she is a pioneer in her field. She's a good role model."

Didn't Christa fit that definition?

"My gosh," she said, "I haven't done anything. Ask me after I've flown."

Christa flew from Connecticut to Prince George's County, Maryland, and addressed twelve thousand employees of her former school district in a giant back-to-school assembly at the Capital Center sports arena. She told them she was their representative for the next year. She said people across the country would judge teachers by her performance, and she promised to live by the rules she set for her students.

"I ask them to be themselves and to do the best they can," Christa said. "I figure if I follow my own advice, I'll represent you well."

Afterward she stopped at Bowie State College, where she was saluted as the school's "most famous alumnus."

"What an exciting time to be alive," she told teachers and administrators at a reception in her honor.

She had invited Tom Campion, her former classmate at Bowie and teaching colleague at Thomas Johnson Junior High, to join her, and Campion had brought his five-year-old son, Jason, who waited for a crush of autograph hunters to thin so he could ask Christa a question.

"Will you meet any little green people like ET in outer space?" he asked.

"I don't know," Christa said, "but if I do, I'll take a picture and bring it back for you."

That night she left for Concord, eager to complete some personal business before she left for training. Finding child care for Scott and Caroline was no problem. Jane Cogswell, the wife of their school principal, agreed to pick them up after school and watch them until Steve finished work, whenever that might be. The Cogswells lived close to the school, the park and Christa's home on The Hill, and, best of all, they were close friends.

Conveniently, so was Eileen O'Hara, her replacement at Concord High. O'Hara had attended Christa's party the night she was selected, and when Charles Foley called to congratulate her, O'Hara had jumped to the phone and asked for the job. She was as familiar as anyone with Christa's course on the American woman, and she followed the same teaching philosophy. She also agreed to assume most of Christa's extracurricular activities — an independent study program, the youth and government program, the Washington close-up program, the New Hampshire close-up program, the New Hampshire Bar Association's adopt-a-classroom program. O'Hara was vacationing in London when Foley called a week later to offer her the job.

The position was temporary, he said, but the obvious reward was "the allure of being known as the teacher who replaced the first teacher in space."

Christa helped prepare O'Hara to replace her, then she helped to find a substitute teacher for her Sunday school class. She asked for a leave of absence from the Junior Service League, withdrew from the Court Jesters, her team in the Concord Co-ed Volleyball League, and told her foursome at the racquet club of Concord that they were now a threesome. One night she bought Christmas gifts for friends and relatives in California, Utah and Hawaii.

But mostly Christa prepared her family for the long separation. She made sure Scott and Caroline had enough back-to-school clothes and that Steve had enough shirts. She bought a microwave oven and stocked up on groceries, including an abundant supply of cornflakes. She balanced the checkbook and briefed Steve on the daily routines for breakfast, baths and bedtimes.

"I have to take care of the little mundane things," she said, "because I'll be a much more alert and interested person in Houston if I know everything is all right at home. If it's not, then I've got a problem. I mean, I can't teach if things aren't all right at my house. That's just my priority."

It was Monday, September 2, her thirty-seventh birthday. It was also Labor Day, or as Steve knew it, Junk Day, the occasion each year when he and two friends, Bill Glahn and Mike Callahan, squeezed their children into the back of his Volkswagen bus and criss-crossed the state treating them one last time to the commercial joys of summer — water slides, miniature golf, carnival rides, lake-front arcades, fried clams, soft ice cream, pizza, an Alka-Seltzer salesman's dream.

Swept up by Junk Day, Steve forgot Christa's birthday, so she was especially grateful the next day when she was showered with going-away presents in an opening assembly for the Concord school district's teachers.

The union gave her a gold charm bracelet. The school board gave her a leather journal with the inscription "Reach for the stars, S. Christa McAuliffe." The superintendent's staff gave her a large globe on a pedestal, "so you don't get lost in space."

"But will all the countries be the same color?" she wanted to know.

Nine other teachers were honored for educational achievements, but only Christa received a standing ovation. She spoke briefly, teary-eyed at times, and explained that it was the first September since she had entered kindergarten that she would not return to school. She said it was for a good reason.

"Yes, I'm going to be getting a new perspective, and yes, I'm going to be getting a wonderful ride in the space shuttle, and yes, I'll be traveling around the country, but the reason I was chosen was that I was a teacher, and because teachers were being recognized nationally as good communicators and as people everyone could relate to. That is very important to all of us."

Most of the teachers knew her reputation as the school's field trip champion, and they laughed when she told them the title of her first space lesson, "The Ultimate Field Trip." They laughed louder when she pointed skyward and said, "Look, there goes Christa Mc-Cloud!"

She closed her speech by describing a gift that Carol Jensen, a high school English teacher, had given her. Jensen had seen someone wearing a T-shirt she liked on a beach in Florida, and she had bought one for Christa. The T-shirt said, "I Touch the Future — I Teach."

"I really appreciate that sentiment," Christa said, "and I'm taking it with me" on the shuttle.

On her way out of the auditorium Christa signed an autograph for Harvey Smith, a reading specialist at the high school.

"To Harvey," she wrote on his class list. "Isn't this a great shot in the arm for education?"

She saw many of the teachers again that night at a reception for her in the governor's mansion. They drank wine and mingled with the governor, a couple of state supreme court justices and members of the state school board. They smiled for the television cameras and mentioned how wonderfully strange it felt to be honored in a state where their salaries, when adjusted for inflation, had actually decreased in the last ten years. Christa's victory was their victory, and they were proud to share her glory.

"From that day on I saw a whole lot more unity and pride among our teachers," said Rosemary Duggan, the principal of the Walker Elementary School in Concord. "It led to some terrific teaching, and in that sense Christa's mission had already been a success."

The shuttle *Discovery* had landed safely earlier that day at Edwards Air Force Base in California. It reminded Christa of *Challenger*'s last landing at Edwards, when its coat of tiles had been damaged in a hailstorm, and it was still on her mind the next morning when she appeared on a local television show after dropping off Scott and Caroline on their first day of school.

"It seems like the shuttle shouldn't be that fragile," she said, "but it is."

She wondered if young people understood the dangers.

"When you think of our children who have not seen the first rockets go up and have not seen the agonizing moments, you know, of 'Is it going to make it?' you realize they just take it for granted," she said. "It's really a new frontier."

Five days remained until training, and Christa had a promise to keep — a final good-bye to the twelve hundred students at Concord High School. Christa spoke from a wooden stage in the school auditorium to the sophomore class on Wednesday, the juniors on Thursday and the seniors on Friday, shadowed daily by a *Life* magazine photographer. The seniors were the hardest to leave, she said, "because I'll never have the opportunity to teach them again. Some of them, I'll never see again."

They filled the auditorium, some of them in Oxford shirts, others in torn T-shirts, a few in Crimson Tide football jerseys for their game that night. Christa wore the lucky yellow jacket she had worn at the White House, and she sat in the front row next to her father as principal Charles Foley, a short, balding man with a stern face and a pleasant disposition, strode to the microphone and welcomed the students back to school.

"For better or worse," he said, "you are the class of '86."

Then, smiling, he introduced the teacher who would make their senior year more memorable than they imagined: "Sharon . . . excuse me, Christa McAuliffe."

The students stood and cheered as Christa bounded up the stairs and onto the stage. She told them she was sorry she would miss their senior year, but she promised to speak at their graduation in the spring and return to her classroom the next fall.

"If the teacher in space doesn't come back to teaching," she said, "something's wrong. I'll be back in '86, so tell your brothers and sisters to take my classes."

Under the glare of the television lights, Christa told them she would take a tiny replica of the school flag with her into orbit. Then she explained she was not riding *Challenger* for a thrill. She said she was going into space to teach them a lesson, and she urged them to consider it carefully in the next few months.

"This is a real special year for you because it's your last year in high school and it's the beginning of what you're going to be doing in life," she said. "You're going to be looking at all sorts of career opportunities and you're going to be looking at colleges and you're going to be saying, 'What am I gonna do? What do I want to do?' Well, reach for it. You know, go for it, push yourself as far as you can, because if I can get this far, you can do it, too."

As they bounced to their feet for another ovation, class president Carina Dolcino led Christa backstage and told her to cover her eyes. The curtains opened a minute later and Christa uncovered them to find the seniors holding three huge banners, one from the balcony, that said, GOOD LUCK FROM THE CLASS OF '86! . . . MRS. MCAULIFFE . . . HAVE A BLAST!

Christa threw her arms around Dolcino to the delight of the *Life* photographer who stood on a ladder behind them. Then, her eyes welling with tears, she held up the tiny Concord High School flag.

"I promise you this will go on the shuttle," she said, her chin trembling. "Have a wonderful year. I'm going to miss you all."

The applause had hardly subsided when Assistant Principal Mark Roth stepped to the microphone. "This is great," he said. "A woman gets chosen, one of eleven thousand, to go into space, and I have to follow her to talk about smoking."

He explained the school's smoking policy as Christa collected her father and quietly left. In forty-eight hours Christa would begin her sabbatical in the space program.

"How's it all going to change you?" a television reporter asked her in the lobby.

"Oh, I have a feeling I'll have my feet on the ground when I come back," she said.

Late that afternoon she asked Steve to meet her at the babysitter's house so he would know where to find Scott and Caroline on Monday. She hosted a backyard barbecue for friends and relatives on Saturday, then walked out the front door on Sunday, ready for training but not quite ready to leave her family. She wished there was an easy way to say goodbye.

Scott was confused. He wore Christa's oversized space camp cap, filled his room with models of the shuttle and enjoyed the attention he received from his classmates because his mother was an astronaut. He was proud of her, but he sure didn't want her to go away. She wouldn't even be there for his birthday in four days. He stood alone and threw rocks at a wooden fence as his father shoved her suitcases into the car.

Caroline had a harder time coping. She had cried when Christa took her students to Washington in recent years and when she had gone alone to Spain the summer before. Caroline had cried each time she had gone away since the White House ceremony, and the confusion had seemed to mount. When Steve took Caroline to the airport to pick up Christa after a recent absence, a woman had asked her who she was waiting for.

"My mommy," Caroline said.

"Is this the first time she's been away?" the woman wanted to know.

"No," Caroline said. "She does this a lot. She comes and goes and comes and goes and comes and goes . . . "

"Is it her job?" the woman asked.

"No," she said. "She just wants to go into space or something."

Now her mother was going away for weeks instead of days, and Caroline, wearing a pink jumper and clutching her stuffed koala bear, wanted her to stay.

"I have one unhappy little girl here," Christa said, holding Caroline in her arms.

When Steve had finished packing, Christa walked across the yard, hugged Scott and kissed him good-bye. Then she kneeled and hugged Caroline.

"Be a good girl," she said. "I love you."

As they hugged each other tighter, their eyes filled with tears and Christa rubbed Caroline's back.

"Okay, honey," she said after a few seconds, "I have to go now. I'll see you later, alligator."

CHAPTER EIGHT

At T-minus ten seconds, her stomach started to tingle. She flexed her shoulders against her safety harness and squinted out the cockpit window. The countdown crackled inside her helmet. At T-minus six, the engines howled and the flight deck shuddered beneath her. She clutched her seat.

The vehicle jolted at lift-off and rose so quickly that within seconds the seventeen-story launch tower was a tiny Erector set in the fiery, fading distance. Arcing above the Atlantic, the ship shook wildly, the engines wailed and the blue sky soon turned black, then gold for an instant as the booster rockets burst free and Christa and the crew climbed toward a starry orbit at twenty times the speed of sound.

The ride felt more promising once the rockets had separated. The vibrations ebbed and the vehicle continued to gain speed and altitude, its velocity flashing on a control panel in the cockpit — 15,000 miles an hour, 16,000, 17,000. They were 70 miles high, not a millisecond off schedule, and then, without warning, they were thrust forward against their harnesses in a sudden loss of power. The engines had failed on the threshold of space.

NASA's shuttle mission simulator had struck again.

Bells rang, whistles sounded and Christa watched the crew tackle the mock crisis, one of dozens they confronted in the $100 million centerpiece of the astronaut training complex at the Johnson Space Center. Engine failure one day, a computer glitch the next, flight control problems after that; the simulator never gave the crew of mission 51-L an easy ride. They entered it to heed NASA's warning that "flying is like parachute jumping: you have to do it right the first time." They tried to prevent what the astronauts described as "not a good day" — a disaster.

The simulator was a computer-driven replica of the shuttle's cockpit, aglitter with thirteen hundred switches that Commander Scobee told Christa "in serious jest" never to touch. It had no middeck, where she would sit on the actual flight, so at lift-off Christa sat behind Scobee and watched images of an actual flight flash by the cockpit windows. She heard the scream of the engines and felt the bone-rattling force of the launch. When the bells rang, she steeled herself for crises that might await her in January. She prepared for almost every kind of crisis but one.

"We never simulated an unsurvivable failure," said Michelle Brekke, a flight director in charge of Christa's training program. "Nobody even talked about it. It's something you just didn't do."

The idea was to train the crew to cope with crises they *could* solve. The mission was to last six days and thirty-four minutes, and bells were bound to ring sometime. NASA needed no surprises.

"We can't afford to have anyone panic," a spokesman said. "One person's panic could endanger the whole crew."

So Christa entered a crash course on living in space and confronting emergencies that might never occur but were conceivable enough that a life insurance company canceled a policy she had bought from Steve's cousin. She arrived on a Monday morning, September 9, 1985, equipped with a new spiral notebook.

"I still can't believe they're actually going to let me go up in the shuttle," she said, smiling, her backup, Barbara Morgan, standing in her shadow as they pinned on NASA identification badges before a half moon of television cameras.

They were to be trained as payload specialists, shuttle guests who had thus far fallen into three categories: engineers for private companies with cargo aboard the shuttle, foreign dignitaries and American politicians. Their formal training was to last 114 hours, about the time it took a student to complete Christa's women's history course and a tiny fraction of the time it took an astronaut to prepare for space flight. They would spend twelve hours in the shuttle mission simulator, fly in supersonic jets and ride the vomit comet, but mostly they would learn the basics — how to work, eat, sleep and go to the bathroom in space.

"The big thing is to make sure this isn't the worst camping trip of all time," said Frank Hughes, NASA's chief of flight training.

The teachers chose wardrobes before they trained, television cameras whirring as they were fitted for sky-blue flight suits, sturdy black boots for the launch and leather-soled woolen moccasins for flailing about in weightlessness. They picked up their red, white and blue helmets and tried on the inflatable rubber pants that would keep their blood circulating during the plunge back to Earth. They selected personal items that ranged from a toothbrush and skin cream to a Swiss Army knife and a sleeping mask. Occasionally, they scurried for privacy.

"Okay, I'm going to the bathroom now," Christa announced to the crush of photographers. "Nobody follow."

"Don't worry, ma'am," one of them said when she returned. "If you're used to working with little schoolchildren, you should have no problem with us."

"Actually, I work with eleventh and twelfth graders," she said with a smile, "but you're right."

On their second day in training, the teachers critiqued NASA's space cuisine, sampling some of the 140 food and beverage items Christa could order for her orbital meals. They graded them from one to ten, with ten at the top of the scale. Christa gave a five to a powdered strawberry breakfast drink, a seven to the rice pilaf, an eight to scrambled eggs (even though they "didn't quite taste like egg"), a nine to Texas barbecued beef and a "nine-plus" to her favorite — broccoli and cheese. She burned her tongue on the chicken consommé, but gave it a nine when Dr. Charles Bourland, the manager of the astronauts' food service, promised it would be cooler on the shuttle. In fact, Christa gave none of the forty samples less than a five, admitting that "after cafeteria food, everything tastes terrific."

Later, she met her crew mates, who gave her a shiny red apple and turned out to be just who Alan Ladwig, the manager of the Space Flight Participant Program, had been looking for when he asked that they accompany the first teacher into space. There was Scobee, of course, a teacher's husband, the enlisted man who had used education as a springboard to success. There was Mike Smith, who had a master's degree in science and three children in Houston's public schools; Judy Resnik, a doctor of electrical engineering; Ellison Onizuka, an aerospace engineer who was the astronauts' liaison to NASA's student involvement program and served in the parent volunteer program at his daughter's school; and Ronald McNair, who spoke louder and more often than any of them about the power of knowledge. (Gregory Jarvis had yet to join the crew.)

A schoolteacher's son, McNair had studied his way from the cotton fields of a small town in South Carolina to the cosmic frontier. McNair, the second black American to reach space, had graduated from a segregated high school at the age of sixteen, earned a doctorate of philosophy in physics from the Massachusetts Institute of Technology and studied laser physics in France. He had married a teacher and spent his adult life urging young people to do what he had done, to push themselves.

"Yes, I did go to the very school you are now attending," McNair wrote to students at his former elementary school after his first space flight in 1984. "Yes, it is true, astronauts are usually from New York, Los Angeles, Philadelphia and Boston. But let the fact that one of them is from LAKE CITY, S.C., serve as a lesson to you that it doesn't matter where you come from, who your relatives are, how much money you have or who you are. Whether or not you reach your goals in life depends entirely on how well you prepare for them and how badly you want them."

The moment he chose to leave his small world and compete against the best and the brightest at MIT, he said, he became a winner.

"Not because I went to MIT," he told students there in 1984. "Many people go to MIT who are not winners. Not because I finished MIT. Many people finish MIT who are not winners. I became a winner because I was willing to hang it over the edge. The unknown was mysterious. The unknown is frightening. You can only become a winner if you are willing to walk over the edge and just dangle it, just a little bit. That's why I say to you students . . . be a winner. Hang it over the edge."

Even after the thrill of space flight ("God, I felt like I was born there"), McNair had considered teaching at the University of South Carolina, and on his return to Massachusetts he had urged the state legislature to increase teachers' salaries. Christa saw a kindred spirit in Ron McNair.

"He's a one-man teachers' union," she said.

McNair's personality reflected his interests. A fifth degree black belt karate champion, he had won an Amateur Athletic Union gold medal, taught martial arts for several years, even studied their scientific foundations as a physicist at the Hughes Research Laboratories in California. He also played the saxophone in the Contra Band, an eighteen-piece group of space center employees, and in jam sessions at Houston clubs. In training, he was the cool jazzman with the quiet control of a karate champion. His conversation was a product of his curiosity; idle chatter bored him. Questions came easily. He asked Christa about her home, her family, her students, her dreams.

"Where do you see yourself in ten years?" he asked her one day.

"I guess I see myself in New Hampshire and in education, maybe in administration or curriculum development," she answered. "I want to have a bigger impact on the system and how it works."

"Do you think you could make a bigger impact than you'll make with this mission?"

"I'm not sure," Christa said, "but I can't wait to find out."

Although Scobee had already given Christa the crew's blessing, her other crew mates punctuated it soon after she arrived for training. To her surprise, they asked her to leave the room one day before a photo session. She returned to find them wearing shorts and short-sleeved shirts, white athletic socks and black mortarboards with dangling tassels. They held apples and Cabbage Patch Kid lunch boxes. Resnik had a child's purse on her shoulder, and they had brought Christa's teddy bear, Radar, given to her by the producers of *Sesame Street*. They put a NASA cap on Radar, gathered around Christa and smiled for the cameras. Christa laughed. They liked her, and she knew it. She had said from the start that she would never pretend to be an astronaut. She had told them she had won the ride of a lifetime and that she intended to go as a team player. Now the crew had accepted her, and she had never been prouder.

"Smile, Radar," she said, laughing, as the photographer focused.

Christa felt at home, at least at the space center. Beyond the security gates and the barbed-wire fences, however, she was misplaced, a native New Englander expecting autumn to deliver the glory of the dappled leaves and the cool Canadian winds, not the throbbing heat and humidity of the drab Texas Gulf. She settled into a furnished apartment at Peach-tree Lane, a sprawling, adults-only complex tucked behind a half-occupied shopping mall a mile from the space center on NASA Road 1. Palm trees grew in the parking lot. Across the street were a Burger King and the Putt-Putt miniature golf course and batting cages.

For about $1,000 a month, her apartment came with tight security, weekly fumigation to fight the creatures that seemed to crawl in at will and air-conditioning she turned low enough so she could wear a sweater and jeans and imagine herself in Concord. She found no solution to the humidity, however. Her first week there she tried to dry her laundry on the dining-room table. The clothes stayed damp for days.

"Want a free perm?" she asked a friend in Concord. "C'mon down. Your hair frizzes before you walk out the door in the morning."

Christa's building, one of about thirty in the complex, overlooked an outdoor pool and a picnic area. She had a bedroom, a living room, a galley kitchen and a dining room she had turned into a study. She often sat at her desk past midnight poring through training manuals, answering letters and writing college recommendations for students at Concord High. She decorated the walls with pictures of Steve and the kids, and she called them nightly.

"Mommy, are you in space yet?" Caroline asked her once.

"Not yet, honey," she said. "I still have some homework to do."

Christa jogged or lay by the pool and studied her workbooks in her free time. Occasionally she read *Good Housekeeping*, worked on her needlepoint or hemmed curtains for her children's playroom at home. Once she baked a pie for the crew with apples she and Caroline had picked in New Hampshire. When Mike Smith raved about it, she baked another one. Christa rarely entertained, though, partly because the apartment was in disarray ("I'm not the neatest housekeeper, you know") and partly because it was her sanctuary. She had sacrificed so much of her privacy that she wanted to protect what remained.

Morgan, who lived two doors down, was welcome. So was Christa's publicist, Linda Long, and Jarvis, who moved into the complex later. The Scobees came once for dinner, but otherwise Christa was alone, free from reporters, hucksters and hangers-on. Only close friends and relatives knew her unlisted phone number, and the manager of the complex shielded her so fiercely that when a local Macy's department store called to verify her address for a credit reference, he said no one named McAuliffe lived there.

"It took me a couple of days to straighten that one out," she said, giggling.

Not even the *New York Times* had the clout to get past her door. When one of its photographers asked to take pictures of her cooking dinner for a journal he was preparing of her year in the space program, Christa politely declined. Later, she compromised and offered to cook a dish of barbecued shrimp in Long's apartment, which was similar to hers. The photographer accepted the invitation, and the *Times Sunday Magazine* ran a picture of Christa and Morgan serving the dinner — complete with wild rice, garlic bread, Caesar's salad and wine — next to a picture of Steve, Scott and Caroline eating a pepperoni pizza as they watched television at home.

Even Steve had to laugh. Here he was, a successful trial lawyer, kneeling at a coffee table and burning his tongue on a meal from Pizza Wheels, a mug of beer in front of him, his two children by his side and his wife smiling over a dish of barbecued shrimp two thousand miles away, all of it right there for the world to see.

"When [the magazine] *Working Mother* is looking for a husband of the year," Steve said, "they know where to find me."

Had the magazine inquired, though, it might have kept its trophy. With little prodding, Steve admitted he had spent his fifteen years of marriage as a fugitive from domestic life. Sure, Christa had gone away for a week or two here or there, but Steve had always relied on his mother or her mother to help out. And even when Christa was there, he had neither the time nor the inclination for housework, a shortcoming she had recently pointed out to him in what he described as "one of those let's-see-if-we-can-cause-trouble-in-the-family tests."

She had spotted it in a magazine as they watched television one night in the family room. The idea was to see which one of them contributed the most around the house. The questions ranged from dishwashing to garbage duty, and Steve sensed defeat from the start. He balked, then agreed to participate when Christa offered to give him the benefit of the doubt on every question. The score surprised neither of them: Christa 94, Steve 6.

Husband of the year, indeed. The day after Christa left for Houston, Steve dropped Home Box Office from his cable television service and replaced it with Sports Channel. He was ready to sit back, relax and root for the Boston Celtics. But within days, he realized that only a housekeeper who came two mornings a week stood between him and domestic disaster. His time to watch sports dwindled, and he began counting the days until Christa came home and he could "bounce back to my old score."

"When I first realized I was going to do this, my attitude was, 'What's the big deal,' " he said one night when the kids were in bed and Christa was in Houston. "But now, wow, it's a different story. There's a lot more to do than I ever dreamed. I'm starting to appreciate why she was never able to keep up with it unless she worked till ten o'clock."

All of it was painfully new to him. After spending his life "barely getting myself out of the house on time," Steve faced his greatest challenges in the morning: making sure Scott and Caroline woke on time, had school clothes to wear, ate breakfast, washed their hands and faces, brushed their teeth, had lunch in their lunch boxes and the proper shoes on their feet. When he learned that Caroline was fussy about her clothes, he allowed her to wear what she wanted to wear — a gray sweat suit three times a week. But Steve was rarely able to satisfy Scott's compulsive need for punctuality, and he often arrived after the bell rang.

"It's a family tradition," Steve said, recalling the morning Christa walked in late to Sister Seretina's homeroom.

Steve discovered as the weeks passed that he was better at getting Scott and Caroline *in* bed than *out* of bed, better at washing laundry than folding it and better at cooking for the kids than cooking for himself. He ate cornflakes three or four nights a week while Scott and Caroline usually ate something he had pulled from the microwave.

"There's nothing in the frozen food section that can be cooked in a microwave that we haven't eaten," he said. "That includes breakfast."

If Caroline complained about the choice, Steve tried to convince her it could be worse, a strategy that backfired one night when a television crew from WBZ in Boston filmed them in the kitchen.

"Daddy," Caroline asked, unwrapping a frozen dinner, "is this really horse meat?"

"Absolutely not," he said, steering her in front of the camera. "Now repeat after me: 'It's turkey, Mommy. Don't worry.'"

Steve's struggle appealed to television people, especially those who had seen his lighter side. Many of them asked, but few were chosen to see him ply his culinary skills. One was Martha Cusick of WNHT in Concord, who found him in the kitchen, clutching a bottle with an unfamiliar shape.

"What's that?" she wanted to know.

"Paul Newman salad dressing," Steve said as the camera zoomed in. "It's good. Have you tried it?"

"No. Is it expensive?"

He checked the price.

"I don't know," he said. "A dollar nineteen. Is that expensive?"

Caroline stood nearby on a chair by the sink and made the salad. Her two Cabbage Patch Kids sat on the counter, but Scott, who was still sick of the press, had retreated into seclusion.

"Put a lot of tomatoes in," Steve said.

"Bunky doesn't like tomatoes."

The salad, as it turned out, was the most challenging course of the evening. The entree was spaghetti with a meat sauce that had been cooked by Margaret Lind, the wife of Christa's lawyer and one of many friends and neighbors who stopped by through the months with home-cooked meals. Steve's job was to cook the spaghetti, and as he snapped a fistful in half and dropped it into a steaming pot, Caroline, elbow deep in salad, interrupted him.

"Can I do some, too?" she asked.

He carried some toward her, but it slipped from his hand and splintered on the floor.

"Oh, Da-a-a-a-addy!" Caroline squealed, scolding him.

He finally brought her to the stove and placed some in her little fingers, coaching her as she struggled to snap them.

"C'mon, strong, strong!" he said. "Break, break! There you go!"

Steve shoved some garlic bread into the microwave and then inspected the salad. There were tomatoes aplenty, all right, but Caroline had buried them under the lettuce. He handed her a spoon.

"Okay, let's stir it up very caref — "

Before the spoon reached the salad, the

spoon fell and clanged on the floor.

"Oops," she said, raising her finger to her lips.

"Well, sometimes we're cooks," Steve said.

"And sometimes we're not," Caroline said.

"When's that?" Cusick asked her.

"When Mr. Microwave cooks!" Caroline said, as if Cusick should have known.

Just then Scott arrived to inspect the salad for himself. He knew Caroline had made it.

"What in the world is it?" he asked, twisting up his face.

"Salad," she told him proudly.

"It doesn't look like it."

"It's salad," she assured him. "It's really salad."

"Well, it doesn't look like the usual everyday salad."

"What's wrong with it?" Steve asked.

"It doesn't look good."

"What d'ya mean? Caroline made it?"

"Oh . . . that's why."

"O-o-o-o-h," Steve said, consoling Caroline. "I think he's just kidding."

When dinner was ready, the three of them climbed a flight of stairs to the family room, where a video library of Christa's public appearances stood next to stacks of magazines and newspaper clippings that grew larger by the day. They ate at the coffee table and watched a Disney movie about Davy Crockett. Steve twirled the spaghetti onto his fork and looked at the kids. It was hard for him to keep them happy, run the house and sleep alone every night in the loft bedroom with the big windows that he and Christa had spent months refurbishing. But he never fussed about the sacrifices. His wife, as hard as it was to believe, was closing in on the ultimate flying experience.

"You know, you're talking about a human being breaking free of the bounds of gravity, orbiting the Earth," he told an interviewer. "There aren't very many human beings who have done that. So I think most of us feel that whatever the price of readjustments, of my taking on things that I probably should have been doing before and hadn't done, whatever those prices are, they pale in comparison to the opportunity."

And Christa wanted dearly to do it right. She knew she represented not only her family, her hometown, her state and the nation's two million teachers, but also every other ordinary soul who had dreamed of space flight. While her five crew mates shared an office in the astronauts' building, Christa went to work in another building, seated across from Morgan at a steel-gray government desk near a hat rack that held a bonnet like the ones the pioneer women wore on the western frontier. A large map of the world hung on one wall, and scrawled on the blackboard was a phrase Terri Rosenblatt had used to relax the ten finalists during their testing and training: "Are we having fun yet?" Radar the teddy bear sat in the corner.

As if she were giving a classroom test, Christa sat with her back to the wall, sunlight filtering through the venetian blinds and reflecting on snapshots of Scott and Caroline she had posted on the wall behind her. Next to the snapshots was Caroline's crayon portrait of Christa in a space suit. Several editorial cartoons that had appeared after she was selected for the flight hung nearby. In one of them, Christa stands at the head of a shuttle classroom, the crew seated at desks in front of her.

"Good morning, crew," she says.

"Good morning, Mrs. McAuliffe," they reply in unison.

In another, a cartoon shuttle circles the Earth.

"On behalf of the entire crew," the commander says, "I want to tell you how nice it is to have a teacher on board, Mrs. McAuliffe.

Well, time to dialogue with Mission Control." To which she says, "Dialogue is not a verb."

From the start, Christa refused to learn NASA-ese ("Imagine the first ordinary citizen in space talking like a robot"), and she consulted her acronym dictionary each Friday when her schedule arrived from the training office. She discovered that CCTV Ops were closed circuit television operations, PHO 35 EQ was a review of the shuttle's 35-mm cameras and WCS Procs were waste collection system procedures. Christa was relieved and amused to hear a female training instructor describe WCS Procs as "potty training."

About half of her 114 hours of training was book work — lessons on how to read the fifty-pound flight data file; how to enter and exit the shuttle; how to operate the cameras, the galley and the $1 million toilet; how to do everything in a weightless environment from cap her toothpaste to extinguish a fire. When she finished a lesson, she took a computer test. If she passed, she got a reward — a supersonic jet ride; a session in the KC-135 aircraft that created the sensation of weightlessness; a morning in the shuttle mission simulator; most often a day in the shuttle mock-up, an Earth-bound reproduction of the actual orbiter.

The mock-up helped her memorize the floor plan of the flight deck and the middeck beneath it. She learned where to find her color-coded food tray, her clothes locker and her sleeping bag. Christa familiarized herself with the galley and the bathroom. She learned the location and the range of the cabin's mounted cameras. She prepared to turn the middeck into history's first cosmic classroom. But again she collided with NASA's education people.

Christa had two jobs on mission 51-L: to conduct two live television lessons and to videotape six science demonstrations that NASA would edit and distribute to schools across the country after the flight. The assignments seemed simple at first, but they soon grew sticky. First Christa clashed with her educational coordinator, Bob Mayfield, a science teacher from Texas who chafed at working with a social studies teacher from New Hampshire.

"This would be a lot easier if she knew science," he said midway through the training. "We don't speak the same vocabulary, and it doesn't help that she has one kind of accent and I have another."

Worse, the science demonstrations excited Mayfield the most and meant the least to

Christa, who considered her mission sociological rather than scientific.

"This isn't just a science program," she protested. "Yes, a lot of scientists will be involved in the space program of the future, but so will a lot of other people. We'll have space post offices, space restaurants, space camps, space construction, space schools, space everything. Let's look at the bigger picture."

Besides, science baffled her. Christa was preparing to demonstrate the effects of weightlessness on plant growth, simple machines, effervescence, magnetism, chromatography and Newton's laws of gravity, subjects as foreign to her as the Texas weather.

"I've been very lost sometimes," she said several weeks into the training. "I've had to be told and reminded, told and reminded, told and reminded."

Mayfield interpreted her confusion as a lack of commitment, and he resented it. He had already simplified the demonstrations to suit her science skills, and now he feared she would allow the project to fail altogether.

Then she surprised him. She juggled her schedule, devoted more time to learning the demonstrations and finally mastered them. He came to respect her.

"She turned out to be an A-plus student," he said.

But there were other problems. Christa wanted to teach the televised lessons her way, not NASA's. Each lesson would last only twenty minutes as the shuttle passed over Central America and the Caribbean within range of a tracking and data relay satellite (TDRS, pronounced "T-dress") orbiting 23,000 miles above the coast of Brazil, so there was little margin for error. NASA wanted the lessons scripted. Christa refused.

"This isn't a stage play," she said. "Teachers don't need speeches. All they need is a lesson plan and their students. It's worked for ages on Earth, and there's no reason why it shouldn't work up there."

She won the argument.

"Knowing Christa," said Brekke, "she wouldn't have used a script even if they gave her one. When she believed in something, she stuck by it."

The lessons could not be entirely free-form, of course. This was television, after all, and there was a lot to do in little time. Christa spent hours in the shuttle mock-up practicing her field trip through the cabin and her lesson on the value of space flight. Several weeks into the training, she performed her first dress rehearsal.

"*Challenger*, this is Houston," a controller announced. "We have good TDRS television. Go ahead with your first lesson, Christa."

Standing on the mock-up's flight deck, she smiled, a microphone in her hand.

"Good morning, this is Christa McAuliffe, live from the *Challenger*, and I'm going to be taking you through a field trip," she said. "I'm going to start out introducing you to two very important members of the crew. The first one is Commander Scobee, who is sitting to my left, and the second one is Michael Smith. Now Commander Scobee is going to tell you a little bit about flying the orbiter and Pilot Smith is going to be telling you a little bit about the Spock, which is the computer that is used on board."

Christa planned to make up the rest of the words as she went along. Her next stop would be the rear of the flight deck, where she would point out a window to the payload bay, explain that it was big enough to hold a school bus and describe its cargo — a communications satellite, which would have been deployed over Hawaii ten hours into the mission, and an observation satellite, which would have been released to study Halley's comet on the third day of the flight and retrieved on the fifth day.

Then she would float down to the middeck to discuss the bathroom and sleeping bags. Ron McNair would be waiting at the galley to demonstrate how the astronauts injected hot or cold water into plastic containers for their food and drinks. He would sip quickly from one of the drinks, then open a clothing locker so Christa could describe the fish-net restraints that kept the clothes from floating away. Near the end of the lesson she would float to the treadmill and talk about the need to exercise in space. As she spoke the rest of the crew would float to the middeck, circle her and hold hands while she assembled a model space station and described the next generation of space flight.

One orbit later — about ninety minutes — Mission Control would cue her second lesson, "Where We've Been, Where We're Going and Why."

"*Challenger*, this is Houston," the controller said in rehearsal. "We have good video. Go ahead with your lesson, Christa."

With a model of the *Kitty Hawk* in one hand and a tiny replica of *Challenger* in the other, Christa stood on the middeck before a navy-blue curtain that covered a wall of storage lockers.

"We've come a long way from the Wright

Brothers' plane to the space shuttle," she said, launching a twenty-minute tribute to the space program.

She would explain that as *Challenger* passed over Mexico, an observation satellite NASA had launched nine years earlier was making the first close encounter with Uranus, providing more information about the planet in a day than astronomers had learned in the 205 years since it was discovered. Christa would then discuss the satellite *Challenger*'s crew had deployed to study Halley's comet on its first pass by the Earth in the space age. And she would mention the network of communications satellites that had turned the Earth into a global village.

Unlike the first lesson, the second one would be confined to the middeck. She would illustrate how space benefits manufacturing by shaking a bottle of oil and water, then a bag of marshmallows and M&M's to demonstrate that neither liquids nor solids separate in weightlessness as they do in gravity, allowing at least seventy combinations of metals to be created in space that could not be created on Earth. She would explain that insulin was easier to produce in weightlessness and that space had improved medical research by allowing scientists to develop latex micron beads that measure tiny bacteria or viruses in the bloodstream. Three student experiments were to fly on *Challenger,* she would explain, including one that required twelve white Leghorn eggs and an incubator for a study on the effects of weightlessness on chicken embryos. Its purpose was to help determine how animals, humans among them, reproduced in space. It was sponsored by Kentucky Fried Chicken.

"I'll talk a little about that one," she said. "I think kids can relate a lot better to eggs than to the other experiments," one of which examined crystal growth, the other grain formation and the strength of metals.

Both her lessons would end with a five-minute question-and-answer period with students from Concord High and Morgan's school in Idaho. The details had yet to be completed, but no one expected a problem. In fact, only one education problem lingered.

Without consulting Christa, NASA had hired a curriculum writer to design a sixteen-page booklet that the nation's teachers would use to prepare children for her mission. When Christa received a draft copy in the fall, she bristled not only at the introduction (which described how hard she had worked on it) but also at its style and substance. She corrected it with a red pen, slashing and circling, and scribbling objections in the margins. She said it was condescending to warn teachers to ask permission to draw chalk outlines of the shuttle on the gym floor. She argued that it was insulting to remind them to check the *Reader's Periodical Guide* in the library for further information.

"That's demeaning," Christa stated. "Teachers know those things."

Even after the booklet was revised, she was so unhappy with it that she refused to sign it or present it to the president of the National Education Association in a ceremony NASA had arranged for the press. When Christa held fast, the education people asked Morgan, whose role during *Challenger*'s mission would be to moderate a daily show about the flight on the Public Broadcasting System. She refused as well.

NASA should have known better. The cheerleader and the girl next door had hit it off. In a relationship that easily might have bred jealousy, even contempt, Barbara and Christa had come together. The doctor's daughter who had attended Stanford had found common ground with a woman who had once packaged Twinkies to help pay her way through Framingham State College. If

Barbara tired of walking in Christa's shadow, she never showed it. If she envied Christa's celebrity, she concealed it. If she resented spending a year away from her mountain retreat to train for a flight only Christa would make, she never said so.

The only time Barbara balked at her back-up role came when a photographer for a national magazine asked her to hoist Christa onto her shoulders in the pool at Peachtree Lane, a loyal subject carrying the queen. She politely declined.

"It's not like I'm the runnerup in the Miss America contest," she said.

And reporters searching for conflict were quickly disappointed.

"Let's get it out of the way," Barbara would say, smiling, when an interview started.

"Do I like Christa? Yes."

"Am I jealous of Christa? No."

"Do I wish I was going up instead of her? Sure."

She admitted, however, that they were "two different people from two different backgrounds": Christa the middle-class mother whose tastes sprang from the popular culture of her times; Barbara, whose husband, Clay, was also a doctor's child, the product of a more refined culture in which Bach never took a backseat to the Beatles. They might not have socialized together had they taught at the same school, but in Houston, Christa piqued Barbara's interest in video games, and Barbara took Christa to hear the symphony. Christa made Barbara feel like more of a teammate than a runnerup — asking her advice in training, inviting her along when a reporter asked her to dinner, saying "we" instead of "I" — and Barbara, whose husband stayed with her for all but six weeks of training while he researched a novel in the Amazon jungle, helped to ease Christa's pangs of homesickness.

"I don't know how she stays so upbeat all the time," Christa said. "It must have something to do with teaching second grade. You must need a lot of energy for that."

They learned over espresso at Barbara's apartment that they shared the same teaching philosophy. Just as Christa's students dressed in zoot suits to bring to life part of American culture of the 1920s, Barbara's dressed in traditional Japanese garb, wrote haiku and ate seaweed to learn about Japan. Just as Christa's students visited a woman in prison to understand human frailty — she was serving eight to twelve years for hiding her boyfriend's drugs — Barbara's visited her lakeside home at night to learn astronomy by gazing at the stars through the same telescope she had used as a child. And just as Christa and Barbara urged their students to learn from experience, they did the same, combining a reporter's curiosity with a researcher's tenacity and an explorer's zeal. Christa and Barbara shared a special trait: they were mature women with a child's sense of wonder.

In training, Barbara and Christa quizzed the instructors. They asked television reporters how they did their jobs; they pleaded with a NASA video cameraman to film a theatrical traffic cop so they could show friends back home; they scribbled notes, collected pamphlets, talked to tourists, turned work into play.

"I want to know it all," Barbara said.

They jogged at night through a nature preserve; shopped; drank margaritas in a floating bar on the edge of Nassau Bay. They sampled the city's ethnic restaurants and went to see *Back to the Future.* At bedtime Christa usually slept easily, but Barbara was restless. She dreamed of one day, maybe forty years from now, becoming the first teacher to land on Mars.

"I go to bed trying not to think about the [*Challenger*] mission," she said. "I go to sleep

328

thinking about music, but I wake up thinking about space. I'm definitely living the experience."

Like Christa, nothing about the training thrilled Barbara more than riding on the KC-135 and the supersonic T-38s. Their rides on the KC-135 were no longer for pleasure — they were needed to prepare Christa's lessons and her science demonstrations — but the teachers always found time to play. They breathed deeply, exhaled hard and tried to propel themselves backward like balloons. They walked on the ceiling and flew like Superman. They folded their arms, threw back their heads and performed a weightless Russian Cassack dance. They played schoolyard games.

"The first leapfrog in zero gravity!" Christa shouted, pushing herself off Barbara's back and floating through the plane's long cargo bay. "O-o-o-ne leapfrog!"

"Tw-o-o-o leapfrog!" said Barbara, jumping over Christa.

"Thr-e-e-e leapfrog!" Christa said, giggling and twirling like a slow-motion gymnast toward the rear wall. "Whoa!"

Later, Christa and Barbara prepared for flying the T-38s, the most perilous and exhilarating of their training experiences. Sleek two-seater turbojets with long pointed noses, the T-38s would break the sound barrier and produce the forces of gravity similar to those Christa would feel on *Challenger*'s lift-off. Aerobatic maneuvers would prepare her for the shuttle's rolling onto its back soon after lift-off and for the pitch and yaw of its return to Earth. Four astronauts had died in T-38s, a training instructor explained, so the teachers carefully studied their oxygen masks, parachutes, ejection seats and survival kits. Still, as Barbara tugged on the skullcap she wore under her helmet, she couldn't resist.

"Hey, Rocky and Bullwinkle," she said.

They left Ellington Air Force Base in formation: Barbara with Mike Smith, Christa with Scobee and a special camera the *New York Times* photographer had mounted in the cockpit in front of her. Within seconds, they knifed through the clouds and boomed past the speed of sound, careening above the Gulf of Mexico at more than 1,000 miles an hour. Then, as Scobee and Smith had done often near Houston and occasionally in their hometown air shows, they executed a series of dives, barrel rolls and lazy eights. The teachers reveled in aerobatic splendor, and the *New York Times* camera caught Christa's eyes smiling, framed by her oxygen mask and her helmet as the jet soared sideways against the horizon.

Then Christa heard Scobee say something to Smith, and seconds later the teachers had the controls.

"What do I do?" each asked, frozen for an instant.

"Take the stick," the pilots told them.

"Then what?"

"Anything you want."

So there they were, Mrs. McAuliffe and Mrs. Morgan, skipping school to fly in formation through the lower stratosphere like a couple of Blue Angels, a barrel roll here, a lazy eight there, everywhere the world rushing beneath them while Scobee and Smith leaned back and looked.

"I couldn't believe it," Barbara said later. "I'd never been closer to the controls than the coach section of American Airlines."

"They really trusted us," Christa said. "They showed us we were part of the team."

Weeks of training remained, but Christa's adrenaline was running fast. She felt as if she were ready to board the shuttle tomorrow.

"Are you scared?" a fourth grader from Anchorage wanted to know when Christa and Barbara conducted a satellite broadcast to schools throughout Alaska several days later.

331

"Actually, I'm very excited," Christa said. "I haven't felt frightened. I'm not sure how I'm going to feel when I'm sitting there waiting for takeoff and those solid rocket boosters ignite underneath me and everything starts to shake. But right now I think instead of being apprehensive I'm just very excited about doing it. It's kind of like the first time you go on a carnival ride, and you've just said, 'Oh, I've got enough courage to do this.' You're just really excited about doing it and maybe conquering your fears."

"We had a taste of those feelings," Barbara added, "before we rode on the T-38s."

On October 5, a month into their training, the teachers were taking part in NASA's national education conference at the Jet Propulsion Laboratory in Pasadena, California, when Linda Long received a call from NASA headquarters.

"Are you sitting down?" Ed Campion said.

"They didn't delay it on us, did they?" Long asked.

"No, nothing like that. You have a new crew member."

"Oh, God," she said, "please don't let it be a politician."

It wasn't. It was Greg Jarvis, a satellite engineer for the Hughes Aircraft Company, a friendly, unassuming man who was balding, color blind, a bit chunky in the midsection and wanted only to fly. To Long's relief, the spotlight still belonged to Christa.

Jarvis would have flown already were it not for politics. Like Christa, he had won a contest, beating out six hundred Hughes engineers to become a payload specialist assigned to fly with one of the company's satellites on the shuttle *Discovery* six months earlier. The shuttle and the satellite had left without him, however. He had been bumped from the flight after two months of training by Senator Jake Garn of Utah, the chairman of the subcommittee that oversaw NASA's budget. He had been reassigned to fly with a Hughes satellite on *Columbia* in December, but then he had been bumped by Representative Bill Nelson of Florida, who headed the House subcommittee on space science and whose congressional district included the Kennedy Space Center. Then Jarvis landed on *Challenger*.

Hughes had sent up nineteen satellites on the space shuttle, but none would fly on mission 51-L.

"Greg's on this flight for no specific reason," Michelle Brekke said. "He's flying because we have a commitment to Hughes."

That was fine with Jarvis, who had once choked up while explaining how proud he was to have been chosen for space flight.

"You look at an astronaut, who is just about a perfect human being," he said, "and here you are, your hair falling out, and they call you. It's like a dream."

As Christa soon learned, he was a fitting addition to the teacher's mission, a tireless student who planned to carry banners from his alma maters — Northeastern University and the State University of New York at Buffalo — into space. "Just to keep my mind active," Jarvis had continued to take evening courses at the age of forty-one, and he was working on a thesis for his second master's degree when he arrived in Houston. He had arranged to carry the diploma aboard the shuttle and receive "the first degree conferred in space." Once, before he had decided on an engineering career, Jarvis had even considered teaching history.

Christa liked him. He was an ordinary guy. He drove a rusting 1968 Dodge Dart and spent hours riding his bicycle with his wife, Marcia, along the coast near their home in Hermosa Beach, California. He and Christa visited often at Peachtree Lane. They drank wine and talked about Boston, where Jarvis had lived in a basement apartment on Beacon Street while he went to Northeastern. They talked about ice cream, Christa's penchant for peppermint, Jarvis's addiction to banana splits, and they talked about Christa's family, Steve's struggle as a single parent, Scott's frog collection and Caroline's confusion about Christa's whereabouts. They were delighted to learn that Jarvis had the same birthday as Caroline — August 24 — and sometimes, on a slow training week, they played Trivial Pursuit. Even then, though, they found themselves considering space. Some of the questions were easy: "What did John Young and Bob Crippen take into orbit?" (Answer: the space shuttle *Columbia*); "Who went into space aboard *Freedom 7* on May 5, 1961?" (Answer: Alan B. Shepard); "How many Russians have landed on the moon?" (Answer: zero).

"Who knows," Christa joked, "maybe one of us will end up a trivia question one day."

As a historian, Christa knew she had already attained trivia status, of course, and if the press kept panting after her as it had since she was chosen, she knew she might even become a household name. She was the hottest space story since Sally Ride.

"She's as big as any of our firsts," said Barbara Schwartz, a NASA public affairs officer who worked full-time in Houston to coordinate Christa's media schedule. "I have dozens of people who want to interview her who will never get the chance."

With a push from NASA, the media had stumbled upon a new concept: the teacher as hero. Forget Rambo, the rock 'n' roll idols and the recalcitrant million-dollar athletes. Here was a role model who spoke plainly, showed no pretensions and harbored no personal ambitions beyond meeting the promise she had made to herself and John Kennedy twenty-five years earlier. The publicity could have exposed her as a fraud, but it didn't. It could have tarnished her image, but it didn't. It could have drained her enthusiasm or emptied her words of meaning, but it didn't. Through it all, Christa was one of us. She was going where none of us had been, and all of us, it seemed, wanted to know her better.

"I'm sorry," Schwartz explained on the telephone one day to a frantic Boston reporter. "I'd set something up if I could, but there just is no . . . more . . . time. Believe me, I wish there was. I've had to say the same thing to everyone."

NASA suddenly needed all the favorable attention it could get. Private companies were complaining louder than ever about scheduling delays. Congress was pushing for cuts in the agency's budget. The president's "Star Wars" strategic defense program was under fire. And NASA's chief administrator, James Beggs, was on an unexpected leave of absence to fight charges that he defrauded the Department of Defense several years earlier when he served as an executive vice-president of General Dynamics Corporation.

Christa was the best thing NASA had to offer, but from September to December her time with the press was trimmed to two hours a week. There were a few minutes for *Life, People, Newsweek, Time,* the *New York Times, USA Today,* the Boston and New Hampshire newspapers. There was time for *Hour Magazine, America* and a few more television and radio shows (to Steve's dismay, she never appeared on his personal favorite, *Saturday Night Live*). There were interviews by satellite for nightly news shows across the country. But there was no time for telephone interviews — hundreds had requested them — or more time-consuming projects like live call-in shows and scripted television specials. As hard as they tried, journalists from Japan, Italy, Fiji and India never got a hold of her either.

It wasn't the first time a teacher had been the focus of a flying publicity campaign. In 1927, the Dole Pineapple Company invited thousands of teachers to apply for a seat on one of the first planes to fly from California to Hawaii (only an army plane had accomplished the feat).

Dole lined up fifteen planes to race from San Francisco to Honolulu, with $35,000 going to the winner, and selected Mildred Doran, a twenty-two-year-old elementary school teacher from Flint, Michigan, to ride in one of them. Her flight piqued the interest of newspapers across the country. As it turned out, though, seven of the planes either crashed or developed mechanical problems on the way to San Francisco, and only four of the eight reached Hawaii. Seven people died, including the teacher.

No one who opposed NASA's teacher-in-space project mentioned Mildred Doran, but a couple of critics talked about protecting civilians from similar dangers. John Glenn said the shuttle should be used for basic research, not for providing "cosmic carnival rides" to "the butcher, the baker and the candlestick maker." Wally Schirra, another of the original astronauts, bristled at NASA's "scarf and goggle syndrome."

"The shuttle is not a passenger plane yet," he said. "In fact, we're still learning how to fly the thing."

Otherwise, dissenters were few, among them the magazine *Human Events*, a conservative weekly that complained about Christa instead of the teacher-in-space program. Noting that she called herself a liberal, a feminist who supported ERA, and that she had married an "outspokenly Democratic" husband, the magazine huffed, "Reaganites they are not. . . . Is this truly the most knowledgeable teacher NASA could find?"

But NEA had come around, and so had such opponents as Geraldine Ferraro, the first woman to run for vice president, who had condemned the teacher-in-space project when it was announced several weeks before the 1984 election. Now when people asked Ferraro why Christa would want to fly on the shuttle, she said, "Because she had a unique opportunity and she grabbed for it. I had a unique opportunity, too."

As her celebrity swelled, Christa received letters by the thousand, so many that NASA printed standard letters of reply and ordered an automated pen to keep up with the requests for her autographed picture. Christa insisted on signing each of them, however, and reading every letter that had been written by a child. She added a personal message to each form letter, and above her signature she wrote "Reach for the Stars."

Meanwhile, Long planned for the madness that awaited Christa after the mission. Johnny Carson's people had already begun fighting with David Letterman's people over who would talk to her first. The network morning news shows had already booked Christa for the first day she left her postflight quarantine. There would be a triumphant return to Concord for a parade and a press conference on Steve's birthday, March 3. Then would come a six-month speaking tour starting in San Francisco and stopping in twenty-seven states before Christa returned to her classroom in September. Along the way, she would speak to the graduating classes at Concord High School, Marian High School, Framingham State College and Bowie State College.

"We had enough requests to stay on the road for at least another year," Long said. "I urged her to do it, but she seemed intent on going home."

Christa missed home more than she had imagined. She missed sitting on the church pew she had turned into a couch. She missed serving dinner on the old wooden table she had discovered at a local antique shop. She missed painting and wallpapering with Steve. She missed reading to the kids at bedtime. She missed her students, her colleagues, her desk. She even missed hall duty.

"The only thing I can truly say I don't miss is averaging grades," she said. "I hate averaging grades."

When Christa felt lonely, she sat at her desk, bundled against the chill of her air-conditioning, and wrote letters.

She wrote to Mark Beauvais, the superintendent of the Concord schools, to thank him and his staff for the globe they had given her and to tell them how proud she felt "to be representing the Concord School System." She wrote to Charlie Sposato to thank him for relieving her tension before her interview in Washington; to Kathy Beres to tell her she would visit her after the mission; and to Joanne Brown, who had told Christa "my prayers are with you, be careful," to assure her the space program was safe.

To Carolla Haglund, the Framingham State College professor who had taught her about pioneer women, Christa wrote that it was not easy to be separated from her family, "but we realized that it was a tradeoff for a chance in a lifetime."

Christa mailed an autographed picture of herself to Alan Ladwig and his cat, Apollo. She wrote to Terri Rosenblatt to tell her she was eager to attend her wedding the next spring. She had balloons and flowers delivered to NASA headquarters on Ed Campion's birthday. She sent messages to her meat cutter and her hairdresser, thank-you notes to reporters and a couple of hundred Christmas cards to people from Hawaii to Germany. The cards featured a family portrait that Jim Cole, an Associated Press photographer, had taken the day after she was selected.

Christa even found time to contribute to *The Teacher-in-Space Cookbook*.

"We now have ignition and liftoff," the editor wrote in her introduction to the 188-page book. "Bon Appetit! The Dream Is Alive!"

The class of 51-L contributed most of the recipes, many of which had space titles: Shuttle Salad, Intergalactic Spinach Special, Cloud Nine Cheese Puffs and Rocket Fuel. Several were regional specialties — Navajo Fry Bread, Maine Lobster, Creole Hush Puppies, Roast Buffalo Burgundy — and a few came from VIPs — Ronald Reagan's Favorite Macaroni and Cheese, John and Annie Glenn's Ham Loaf, James Beggs's Chicken Mosel, Jesse Moore's Chicken Jambalaya,

Dick and June Scobee's Peanut Butter M&M Cookies. And then there were Christa's: Spicy Apple Pancakes with Cider Sauce and Broccoli Casserole.

Christa used the phone when writing wasn't enough. She called Caroline's kindergarten teacher several times to check on her progress. She called to express her sympathy to a colleague in the Concord High social studies department whose wife was dying of cancer. She called friends and relatives often, and whenever she got a chance, she went home, several times for a weekend, and once, in October, for a week.

On the October visit, Christa spent a Saturday and Sunday with her family at her parents' house, shielded from the media. On Monday, when Steve went to work and the kids went to school, Christa snuck in the back door at Concord High, sat at her desk and talked to students and teachers about her training — "You'd be so mad if you knew how much fun I was having" — and about what she had missed — Education Secretary William Bennett's visit to her classroom, President Reagan's speech on the state house plaza, the daily media crush her celebrity had caused at the school.

Christa regretted the disruption and deflected some of it on Tuesday by making a television appearance and conducting a half dozen interviews at home. On Wednesday, she and Steve had dinner at the White House. The president was entertaining the prime minister of Singapore and had invited 120 guests, among them Raquel Welch, Sylvester Stallone, Natalie Cole, Shirley Temple Black, Michael J. Fox and the McAuliffes, to join him in the State Dining Room. Steve sat with Raquel Welch's husband at "the table for famous spouses." Christa sat next to the president.

"All those people, and he invited a high school teacher to sit beside him," she said. "I couldn't believe it. I figured I'd be lucky to meet him in the receiving line."

Over Jellied Sole Timbale and Mandarin Sunflower Salad, Christa told the president she was in the midst of the most remarkable year of her life.

"I've had some great experiences before all of this, and I'll have some great ones afterward," she said, "but when I turn ninety and look back, this year will definitely rank in the top ten."

She was proud to represent the teaching profession, she said, and eager to get back to her classroom to share what she had learned.

Reagan told her about his health, his acting career and his ranch in California, and when the main course arrived — Filet of Veal with Ginger Sauce — he told her he wasn't hungry but that he would take two pieces because if he didn't no one else would. When she told Reagan she was sorry she had missed him in Concord, he said nothing, realizing he had blundered by neglecting to mention her on his visit.

The president of Coca-Cola sat on Christa's left, and they had a friendly chat, but Christa was occasionally distracted by passing guests. She had never seen so many celebrities in the same room.

"Raquel Welch is beautiful," she told friends later, "but when you see Sylvester Stallone in person, he doesn't look much like he does in the movies. He's really pretty short."

Christa's vacation was nearly over when she returned to Concord, but her childhood friends from Girl Scout Troop 315 were meeting for the annual reunion at their mountain campground. She packed her pup tent and joined them. She sat by a fire, embers popping in the cold October night, and told them about space food, about the crew picture with the mortarboards and the tassels, about book offers and movie offers, about Steve's escapades as a single parent, about Sylvester Stallone. They laughed into the night, their voices echoing through the pines.

She left in the morning as she was due back in Houston the next day, and there was still so much to do: sew Halloween costumes for the kids, return library books, stock up on groceries, spend some quiet time alone at home. When Linda Long asked her which plane she wanted to take back to Houston — a flight at 5:30 P.M. that would get her to her apartment by 10:00, or a 10:25 flight that wouldn't get her there until after 2:00 A.M. — Christa chose the later one. She wanted to stay and tuck the kids into bed.

As the fourth American mother to fly in space, Christa wanted to prepare her children as well as she could for the fire and thunder of a shuttle launch. She had heard the story of astronaut Anna Fisher's two-year-old daughter, Kristin, who had watched her mother take off from the Kennedy Space Center and screamed in horror, "Oh, Mommy" when the great steam clouds seemed to swallow the orbiter on the pad. She had heard how Fisher's husband, William, also an astronaut, tried in vain to console the little girl. She knew neither of her own children liked loud noises.

346

Caroline turned up her Michael Jackson *Thriller* tape to dance once in a while, but she hated thunder; and Scott was the only kid Christa knew who jumped up to turn down the television when the volume rose during commercials. She wanted them to know what to expect. The problem was that she had never been closer to the space program than a television before she entered the teacher-in-space contest. She had never seen a shuttle launch.

"You know, it would be nice if I saw one at some point," Christa told several NASA officials.

She got her chance the day before Halloween. It was the ninth flight by *Challenger*, the last one before her own, and NASA went out of its way to let people know Christa was there, including an unprecedented visit to the launch pad less than twenty-four hours before the flight. *Challenger* stood in its launch position, its booster rockets loaded, its fuel cells already filled with volatile liquid hydrogen and oxygen, and NASA took two dozen reporters and photographers there after confiscating their matches and cigarette lighters. Never before had the press been permitted to visit the pad so late in the countdown. The reason was Christa.

For the benefit of the photographers, she climbed onto the launch platform and posed, the giant, apricot-colored fuel tank and the pencil-shaped booster rockets rising behind her. She wore a print blouse and plain brown slacks. She had a purse slung over her shoulder, and she looked like any of the thousands of tourists at the sprawling space center. She smiled and waved, and the photographers clicked wildly. Morgan stood in her shadow.

"It's amazing," Christa said. "We've seen the mock-ups and everything, but we're only used to seeing the cargo bay and the orbiter. To see the superstructure and the tanks is unbelievable."

"Has the reality of it hit you yet?" a reporter asked.

"The reality just becomes that much more every time we get a little closer," she said. "I still don't think it's going to hit me until I start walking across that bridge and get into the orbiter when it's about to take off."

Just so there was no misunderstanding, a NASA escort made it clear that the visit to the pad was a one-shot exception to agency policy.

"We never take people up here," the escort said, "but it's good for Christa, and it's good for NASA."

The largest space crew in history — five NASA astronauts and three European scientists — walked across that bridge and climbed into the orbiter the next morning. Flat on their backs, in steel chairs facing the sky, they waited, the countdown crackling in their helmets. Christa remembered what Dick Scobee had said about the wait — "You gotta be dead or crazy not to get excited" — but she had little time to think about it in the final hour of the countdown. Several guests in the viewing area had recognized her, and soon Christa was swarmed by people who wanted to get her autograph or wish her luck or take her picture. Others took turns posing with their arms around her. They kept coming until the countdown dipped below nine minutes and NASA officials shooed them away.

With Morgan at her side, Christa stood before a scoreboard-size digital clock on a field in front of the press center, four miles from the launch pad. To the dismay of the *New York Times* cameraman, who had been promised exclusive pictures, the teachers were surrounded by photographers and reporters. Christa rubbed her hands in anticipation as the voice of a public affairs officer boomed the final seconds of the countdown through the space center's network of loudspeakers.

"You'll feel the sound" [of the launch], a NASA official assured her.

When the rockets ignited, steam clouds billowed from the launch pad and *Challenger* left the ground. So did Christa, hopping up and down, raising her hands above her head and clapping.

"Oh my God!" she shouted. "Look at it!"

The shuttle, trailed by a seven-hundred-foot fire, rose into the blue Florida sky on a ribbon of fierce white vapors.

"It's so quiet," she said. But then came the sound: a hammering series of shock waves that rattled the rib cage and shook the ground, echoing until the booster rockets burned out and exploded away from the shuttle in a final flash of light. Then it was quiet again. *Challenger* had disappeared. Only its towering white tail remained.

"Oh my Lord," she said, clutching Morgan's elbow with one hand and wiping away tears with the other. "It's beautiful!"

The crew had completed what pilot Steven Nagel described as "sort of a ho-hum launch." Soon television sets at the space center showed pictures of the astronauts floating into a portable science laboratory in *Challenger*'s cargo bay, where they would conduct seventy-six experiments on a nearly flawless seven-day mission. Christa was answering questions at the space center's geodesic press dome.

"Oh, it was a very emotional experience," she said of the launch. "It was brighter and more real than I had expected. That light was so bright. It was gorgeous."

"Now how do you think your children will react?"

"Well, I'm still a little worried about the noise, but they're not like me," she said. "They like carnival rides, so I'm sure they'd love this."

A week after the *Challenger* crew had touched down at Edwards Air Force Base, they showed slides of the sights they had seen: the Himalayas, the Alps, a Japanese volcano, the Great Wall of China and the Grand Canyon. They described trick or treating among themselves with Halloween masks made from pages of the flight data file. They talked about "falling in love with floating," and when someone asked how *Challenger* had performed, the commander, Hank Hartsfield, announced that "the bird was in real good shape." A NASA executive said the success of the mission proved that the shuttle program had achieved "maturity."

Four days after landing, *Challenger* rode piggyback on a Boeing 747 cargo jet from

California to the Kennedy Space Center to be prepared for mission 51-L. Christa had returned to training. She would fly in fewer than ninety days, and there was still plenty to do — meetings with astronomers to learn about the stars, meetings with photographers to learn the shuttle's cameras, meetings to review the flight plan, meetings with developers of the space station, meetings to discuss the student experiments, meetings with public affairs officers, meetings with Mayfield, meetings after meetings. Some of them seemed endless.

"We've already been here two hours," Christa complained to an instructor. "Do we have to talk about it for another two hours? Can't we do this any quicker?"

Suspecting she had spooned the same medicine to her own students, she vowed, only half kiddingly, to assign less homework when she returned to her classroom. Still, her enthusiasm impressed her training director.

"She's one of the most inquisitive payload specialists we've had," Brekke said. "She radiates excitement and wonderment."

Despite the danger, Christa relished any hands-on training she could get. One day she donned firefighter's gear with five other payload specialists, including Jarvis, Morgan and two Italians training for a later flight, and doused a small pit fire with a carbon dioxide extinguisher. Then she and Jarvis, both wearing helmets and oxygen masks, used a hose to try to extinguish a larger pit fire that spewed flames several feet high and swallowed them in smoke. Struggling against the pressure in the hose, they surged forward, Christa controlling the nozzle and Jarvis trying to keep them both from falling over.

"What a team," she said, smiling, as they wiped away the soot and sweat afterward. "Thank goodness this is only a field trip."

By the middle of November, they had begun training more often with their five crew mates, each of whom had been chosen for the flight for a specific skill: Scobee for his shuttle experience, Smith for his reputation as a pilot, McNair for the physics background he would bring to the study of Halley's comet, Onizuka for his ability to deploy satellites (he had done so on an earlier military mission) and Resnik for her knowledge of the remote manipulator arm they would use to release the satellites and retrieve the comet watcher. The seven of them had spent so much time training for their individual duties that their time together was important.

Christa cherished that time, the seven of

them clad in flight suits with American flags on one arm, NASA logos on the other and crew patches on their chests. Christa also wore the teacher-in-space symbol, a shuttle soaring toward the stars through the eternal flame of knowledge.

Together they rehearsed the flight in the shuttle mock-up. At liftoff, Resnik and Onizuka sat behind Scobee and Smith on the flight deck while Christa sat next to Jarvis on the middeck. McNair sat around a corner to their left by the circular hatch they used to enter the orbiter. They practiced taking off, entering space and stowing their ninety-pound aluminum seats, which in zero gravity, of course, would weigh nothing. They trained for living on the flight deck, which contained the cockpit and the crew station for the remote manipulator arm, and the middeck, which housed the galley, the toilet, the storage lockers, the waste disposal and the sleeping bags in an area that Frank Hughes said was "about the size of a kitchen in an efficiency apartment." They rehearsed the preparations for reentry, waddling about in pressurized G suits to clear loose objects from the cabin and ensure that no one was struck by them on the high-speed plunge back to Earth. When Jarvis ran out of storage space, he asked Christa for help.

"Do you have an extra storage bag?" he asked.

"Sure," she said. "Teachers are always prepared. Or maybe it's from all my years of Girl Scouting."

A few minutes later, an instructor's voice sounded on the intercom. "That's it," he announced. "We just landed."

Christa applauded.

They were less fortunate on another flight in the mock-up. They crashed into the ocean, an event for which they had prepared by reviewing the shuttle's survival kit, complete with an instruction booklet.

"Pleasure reading," Christa said, laughing.

One by one, they had practiced their escape drills in the mock-up, first shimmying down a cable from an emergency exit on top of the cabin, then leaning out a side exit, grabbing a steel bar and swinging themselves into an imaginary life raft several feet below. Christa had trouble holding on to the bar, and she slipped onto a cushion on the floor before she should have. No one complained. They knew the risks of water landings.

"We don't expect a good day in that case," an instructor said.

The chief of flight training put it more bluntly. "You don't want to go into the water," Frank Hughes said. "This mother is not a boat."

Everyone but Christa and Jarvis had homes and regular routines awaiting them after work. They lived in the suburban subdivisions that had sprung up within five miles of the space center. They had nine children between them — only Resnik was single — and plenty to keep them busy. McNair played his saxophone; taught karate to children, including his three-year-old son, Reggie, at an inner-city church; and attended Bible study classes. Smith jogged through his lakeshore neighborhood, water-skied with his wife and three children and attended Sunday morning Bible classes at a Baptist church a short walk from the space center. Scobee, the oldest crew member at forty-six, walked hand in hand after dinner with his wife, June, in their subdivision near Ellington Air Force Base. With his seventeen-year-old cat, Py, at his feet, he often spent time painting — aircraft were his favorite subjects — or woodworking. On weekends Scobee flew a private stunt plane he owned with James van Hoften, an astronaut who had accompanied him on his first space flight.

Onizuka, the grandson of Japanese immigrants who had worked on sugar plantations in Hawaii, lived less than a mile from Scobee in a modest brown-brick tract house in the Meadowgreen subdivision. It was the first house he had owned, and when he wasn't watching his two daughters play in the local youth soccer leagues or puttering with his car in the driveway, Onizuka was tending a garden in his small backyard. Besides his Asian American features, he hardly looked like the astronauts of Christa's youth. He was short and rather pudgy with a receding hairline and unfashionably long sideburns, and he often wore jeans and white socks to work. Onizuka had a reputation as an exacting engineer, but he was easily one of the most popular astronauts among civilians and fellow members of the corps at the space center. Before each of his shuttle launches, he gave the staff at Mission Control a box of pineapples, papayas and macadamia nuts to enjoy during the flight. He organized parties at Pe-Te's Cajun restaurant, and he hosted a real Hawaiian luau each year in his backyard.

Resnik was the only crew member who was not listed in the telephone book. The astronauts of 51-L were community people, family people, and after work Christa saw them together only when the Scobees invited them to a backyard picnic or, more often, when Onizuka organized field trips to local happy hours.

The Outpost, Pinata's, Joshua's — they visited them all. But Pe-Te's, where pictures of each of the crew members except Jarvis hung on the walls, continued to be a favorite. So did Frenchie's, an Italian restaurant tucked into a tiny shopping center near a Baskin-Robbins and a beauty parlor on NASA Road 1. At Frenchie's they sat amid pictures of shuttle launches and astronauts, their own pictures among them, drank chianti Classico and Cantina Cattavecchi, and ate their favorite meals — Scobee the veal fetuccine, Resnik the scallopini pescatore, Christa a rich tortellini.

"They were so full of life," said the chef, Giuseppe Camera, who called Scobee "El Commandante" and Resnik *"mi bella mora"* beautiful brunette. "You just knew they enjoyed every minute of their time together."

The Scobees befriended Christa and often invited her to dinner. Once she played volleyball with Brekke in a space center league and went out afterward for beer and pizza. Christa spent time with Morgan and Jarvis, and she socialized more and more with the entire crew, but she still missed Steve and the kids. She was delighted when they visited in November.

They stayed several days, and at night they sampled the city's ethnic fare, once eating at a Chinese restaurant where they cooked their own meals at their table. They slept at Peachtree Lane, Christa, Steve and Scott in Christa's apartment, Caroline in Morgan's when Christa ran out of beds. Christa read the children stories and tucked them into bed.

One day they toured the space center.

"See those containers, kids?" Christa said, pointing ahead of her as they walked through the mock-up's giant cargo bay. "They'll have experiments in them on the real flight."

Scott and Caroline were happy to see their mother, but neither the containers in the cargo bay nor much else about her new workplace intrigued them. So what if they were retracing the steps of American space heroes. When Sally Ride first flew on *Challenger*, Scott was six years old and Caroline three. They knew a lot more about R2D2 of *Star Wars* than they knew about Sally Ride of NASA. Besides, the sooner their mother came home from this place the better.

"Mommy, I'm hungry," Scott said, ignoring the containers.

"Me too," Caroline said.

On their way out of the shuttle, they needed to climb through the circular hatch

359

and step over a deep, unprotected shaft to a flight of stairs. Caroline went first. She was uneasy. She looked over her shoulder at Christa.

"It's all right," Christa said. "You don't need any help."

Caroline leaned over the edge and hesitated.

"Go ahead," Christa said in a soft, reassuring voice. "You can do it, brave girl. Go ahead."

She did it, and Christa swept her up in a congratulatory hug.

Before the tour ended, the McAuliffes had seen what the other million and a half visitors to the space center saw each year: moon rocks, films of shuttle flights, photos from space, a replica of the lunar landing module, a giant Apollo Saturn rocket, Mission Control, the gift shop, the cafeteria. They visited training facilities that were off limits to tourists, among them a pool that was twenty-five feet deep with a full-scale shuttle mock-up at the bottom. Christa explained that astronauts training for space walks — no such walks were planned for her flight, but Onizuka and McNair had prepared for them — donned three-hundred-pound space suits and entered the submerged shuttle to simulate weightlessness. She took a picture of Scott and Caroline by the pool. Later, she took them to meet Dick Scobee.

Scobee was a very important man, Christa told the kids. He was the commander, the man in charge of the shuttle *Challenger*. She made him seem so much larger than life, in fact, that Caroline started to have second thoughts about meeting him. She grabbed Steve's leg as they waited outside his office and said, "If he's scary, I'm not going in."

He was not scary, of course, but when Christa asked the kids later what they liked most about their visit to the space center, Caroline thought about it for a minute and said, "The tuna sandwich" at the cafeteria. Nothing had particularly appealed to Scott. The best part of the trip, he said, had been the batting cages across from Christa's condominium.

"In a way, that's good," she said. "Why should kids be so awed by the space program?"

When they returned to New Hampshire, Steve learned that Christa was under investigation by the FBI. At NASA's request, an agent had begun a background check to make sure the first private citizen to enter space had nothing to hide.

"This is the strangest assignment I've ever had," the agent admitted. "It's the first time I've had to check on somebody *after* they've been chosen."

The people he interviewed found it strange for another reason: perhaps they were naïve, but some of them thought he was wasting his time looking for dirt on Christa. Others considered his questions, well . . .

"Real insidious," said Eileen O'Hara. "He wanted to know if Christa had had any affairs, if Steve had had any affairs, things like that. I was appalled."

So appalled that she told him if he wanted answers to his questions he would first have to stand before her class, which would have been Christa's class, and explain what he was doing and why. Before he had finished, three students stood up without prompting and told him what a good teacher Christa was.

When the agent visited Anne Malavich, Christa's best friend since childhood, she took pity on him. Really, she wondered, did anyone expect indiscretions from a woman who played a nun in *The Sound of Music*? A woman who wore her Girl Scout uniform in college? A woman who took her father to her senior prom?

She told the poor guy he was chasing ghosts.

"I know this sounds too good to be true," Malavich said, "but you won't find anything unusual or subversive about Christa McAuliffe. She's just your basic ordinary person who happens to be in extraordinary circumstances."

The FBI agent's appearance at Concord High was another in a series of strange distractions at the school. After the tumult of Christa's departure, Education Secretary Bennett had arrived. At first he had promised to keep the visit low key, but then he had turned it into a small circus by allowing nearly twice as many journalists as students to hear his lecture on "The Federalist Papers." In his diary that night, Mark Beauvais, the superintendent of schools, had written, "It was nice to have [Bennett] leave so we could get back to business as usual."

They didn't. The daily visits continued by reporters counting down the days to the launch, and then on December 3, as Christa answered letters in her office in Houston, came a tragedy that shook the school and the city. Shortly after 8:30 on a Tuesday morning, Tom Herbert, Christa's boss, was taking his law class on a field trip — a drunk driving trial at the county courthouse. As the students waited near the principal's office to leave the building and board a school bus, two of them, including a boy Christa had taught in Sunday school, were taken hostage by a troubled drop-out armed with a double-barreled shotgun.

Louis Cartier had quit school ten days earlier to work at a dry cleaner's with his mother. He had been an outcast at Concord High, a short, slender boy with a mild case of acne and puffy cheeks that prompted his classmates to call him "bubble face" and "chipmunk," a nickname Christa had endured as a child. Cartier was sixteen years old.

Feeling rejected at home as well as at school, he talked excitedly at lunch one day about a movie he had seen the night before. It was *The Day After*, a portrait of a nuclear war between the United States and the Soviet Union. Cartier said he had enjoyed seeing all the people getting blown up and wished that it would "hurry up and happen."

At 7:00 on the morning of December 3, he went to work at the dry cleaner's. At 7:30, his mother saw him talking to the boss and heard the boss tell him, "If you don't feel good, go home."

He did. He grabbed a bottle of wine, strapped an ammunition belt around his waist and pulled the shotgun from his father's gun collection. He walked less than a mile to the high school, stopping for cigarettes at a convenience store on the way.

"Where are you going with the gun?" the store owner asked him.

"Hunting ducks," he replied.

It was not the season for hunting ducks.

Cartier entered the principal's office at about 8:20 and confronted Mark Roth, an assistant principal whose child was a classmate of Scott McAuliffe. For a moment, Roth thought the gun was part of a class demonstration.

"Do you want to leave it here?" Roth asked him.

"Nope," Cartier said, and pointed it at him.

Roth said he would have to call the police. He shut the door on Cartier, who began roaming the halls, where he came upon Herbert's class and took the hostages. He offered them wine, and when they refused he threw the bottle against the wall. The Concord police had arrived by then. Cartier and his hostages stood in a corridor with stairwells on both sides. The police stood in the stairwells, Officer Michael Russell with a handgun drawn in one stairwell, Sgt. John Clark aiming a .33-caliber shotgun from the other side.

Clark was a friend of Christa. He had guarded her against the media crush after her selection, and he had escorted her several times to the airport. He belonged to the Friends program, which matched understanding adults with the city's disadvantaged youths, many of whom lived in Cartier's neighborhood. And he had spoken often to Christa's law classes, telling them how proud he was that he had never fired a weapon in the line of duty. Now he was trying to coax Cartier into dropping the gun. Cartier refused.

"He was terribly impassive," said the principal, Charles Foley, who also tried to reason with him. "When we asked him what was wrong, what he was mad about and what he wanted, he said nothing."

Cartier freed one of the hostages when the boy offered to help him get a ride away from the school. He held the other one, Christa's former student, and refused requests by Foley, Roth and Don LeBrun, the school's football coach, to allow them to trade places with the boy. The standoff continued for about fifteen minutes as students waited in classes throughout the building, some several feet away. Then LeBrun asked Cartier again if he could trade places with the hostage. Cartier responded by aiming the shotgun at him. LeBrun dove to the floor, and Russell fired his revolver, striking Cartier in the ammunition belt.

"Why did you hit me?" Cartier said, firing at Russell but missing him.

An instant later, the hostage fled and Clark fired twice, hitting Cartier once in the wrist and once in the head. The head wound was fatal. Cartier was taken away by ambulance, and the school's twelve hundred students were sent home. The community mental health service was called in the next day to help everyone — students, teachers and administrators — cope with the tragedy.

At no time during the training had Christa felt so far from home. She had called LeBrun and Clark to offer her support. She had called Foley, Roth and Herbert. She had called her friend O'Hara and her superintendent, Beauvais. She had called the boy from her Sunday school class.

A week later, Christa was still upset.

"When I heard what had happened, I was desperate to talk to somebody," she said. "I almost went over to see Barbara's husband, but then I called Steve at the office and of course probably got him all upset. It was frustrating. I just felt like I needed to be at the school to help out the kids. It was like Scott or Caroline getting hurt at home, like Caroline saying she had fallen down or something like that. Even though I probably couldn't have prevented it, I wanted to be there to comfort them."

Through it all, Christa had tried not to let the tragedy disrupt her training. Time was running out.

In early December, Christa left the office she had shared with Morgan and joined Jarvis at the astronaut building in a cubicle down the hall from the rest of the crew. Tradition died hard, and while NASA's astronauts had tolerated sharing the shuttle with guests, they had yet to tolerate sharing their offices. Still, the proximity saved the crew time and trouble in the weeks before the mission. So did their final press conference.

Under NASA policy, each space-bound crew met the press thirty days before they flew. The briefings helped the media learn about the mission and the crew (astronauts like Resnik, who normally declined interviews, could not skip this one), and they allowed the astronauts, most of whom considered reporters a nuisance, to spend the rest of their training in seclusion. Once the briefing had ended, the crew would grant no more interviews until seven days after the flight.

"It'll be nice," Scobee explained, smiling, "to press on without the press."

The briefings lasted two days. On the first day, reporters met the people behind the scenes — the flight director, the experts responsible for the payloads on board, the students who had developed science experiments for the mission. On the second day, they met the crew. They could attend the briefings in Houston or participate by closed-circuit television from NASA centers across the country. Several dozen of them wanted to know about mission 51-L.

The launch was scheduled for January 22, but the briefings were held on December 12 and 13, about ten days earlier than usual, to allow space writers time to cover the countdown to the launch of the shuttle *Columbia* on December 19. Randy Stone, the lead flight director for 51-L, opened the two-day conference in a small, windowless auditorium at the space center's public affairs building. He gave an overview of *Challenger*'s mission, and when he listed the members of the crew, he referred to the teachernaut as "Christa Mc-Coffee."

A reporter asked Stone what he considered the highlight of the mission.

"It really depends on who you talk to," he said. "The communications people are very excited, of course, about getting up this other

TDRS satellite. We've been trying now for two years, and it's a major event because it will replace our network of ground tracking stations around the world. But the astronomers think the Spartan Halley is the most important thing that's ever flown, because this is the only time in their lives they'll get this kind of data. And if you talk to people in the education field, they think the fact that a teacher is flying is the most important thing."

"And you?"

"Personally, I believe the first private citizen in space is historic," he said. "I'm glad it's taking place on a flight I have something to do with."

The reporter then reminded Stone how to pronounce her name.

After a briefing on the TDRS satellite, a public affairs officer announced that NASA was "departing a little bit from procedure." Frank Hughes, the chief of flight training, was about to discuss astronaut training for the newcomers who had begun to cover the space program since Christa's selection.

"I'm going to assume you know nothing about the process," said Hughes, a plain-speaking training veteran who had helped prepare for space flight every NASA astronaut since Alan Shepard. After a half-hour slide show, he fielded questions.

"Does the training provide any psychological preparation?" a reporter wanted to know.

"No," Hughes said. "In the early days it was a big deal, but now the shuttle is an old lady's ride. It's three Gs max going up and two coming home. You can get that at Astroworld," a local amusement park.

"Have you had a chance to judge Christa's strengths and weaknesses?"

"Quite honestly, I would put it this way," he said. "She's going to go up and do some pretty good experiments and everything for us, but this is a lark. It's a big gee-whiz. The training she's been doing is just for habitability, learning all the space food stuff and reading the workbook about the john before she tries it, things like that. We run the astronauts' tails off, you know, and they try to stay awake a couple of extra hours just so they can look out the window. She's going to be unscheduled for a lot of the time on orbit, and that's the most wonderful thing that could happen to her. She can look out the window all she wants. She's going to have the most wonderful time of her life. I hope she gets some great souvenirs out of this."

Hughes had hardly finished before Christa's parents, Grace and Ed Corrigan, arrived for their first visit to the space center. The Corrigans had basked for several months in their daughter's reflected glory. They were a typical suburban couple, a substitute art teacher and a retired accounting manager who suddenly found themselves central figures in one of the happiest news stories of the year. Reporters had wanted to learn more about them. Old friends had called to ask if that was really their daughter who was flying on the shuttle. Former classmates had stood in line to talk to them at Ed's forty-fifth high school reunion. They had pushed aside a family portrait that had hung in their living room for thirteen years to make room for a picture of Christa in her astronaut blues. They had collected a small mountain of news clippings in a corner of her old bedroom, and they had smiled when people drove past their house and said, "Hey, that's where Christa McAuliffe's parents live."

Grace even handled one of Christa's public appearances. It was homecoming weekend at Framingham State College and the theme, in Christa's honor, was, "Homecoming — It's Out of This World." Christa was unable to break away from her training for the weekend, so her mother filled in, riding a Rolls-Royce to the school's football stadium, where she walked to midfield at halftime and crowned the homecoming king and queen. The college president explained that since Christa was busy, the school had done the next best thing by inviting "the woman who brought her into this world."

"It's fun to be the parents of the one who was chosen to be one of eleven thousand," her father said. "It's amazing."

"And the funny thing," her mother said, "is that it's something we knew from the start she could do."

On their first night in Houston, Christa's parents had joined her at a party the wives of the 51-L crew had given for the wives of the crew on the upcoming *Columbia* mission. It was a NASA tradition, and though Christa, Morgan and Resnik were not wives of crew members, they gladly attended. They gathered at June Scobee's house for coffee and pastry, more than a dozen women and Ed Corrigan. He spoke with Sally Ride, whose husband, Steve Hawley, would fly on *Columbia,* and chatted with Resnik. He threw his arms around Morgan when he met her for the first time.

"It's like we have another daughter," Ed gushed.

Now he and his wife were posing for photographers as they waited for Morgan to describe Christa's lessons to the assembled media. Neither of them seemed upset when a public affairs officer introduced them to reporters as "Grace and Ed McAuliffe," and when a reporter asked Ed if he was nervous about Christa's flight, he said, "Next month — I'm saving it." Then he admitted his stomach had turned when he watched a recent launch.

"It must have been because Christa's going up," he said.

Grace was more confident. "I'd change places with her in a minute," she said.

As Morgan described the mission's educational highlights, the Corrigans sat with the reporters. Christa watched on a closed-circuit television in a room down the hall, explaining that "it wouldn't be fair for me to sit in there." She knew the reporters would swarm her and ignore Morgan. Besides, she would have her chance tomorrow.

The first day of briefings ended with the son of a shopkeeper stepping to the podium and putting the *Challenger* mission into perspective. Richard Cavoli, a twenty-one-year-old junior at Union College in Schenectady, New York, had asked Annette Saturnelli, his high school chemistry teacher, to help him enter a national science contest five years earlier. He had won the contest and his experiment, aimed at improving the quality of X rays, had been chosen to fly on mission 51-L.

"There's a special moment, a magic that happens between a student and a teacher," he said. "A student lets his mind take a journey; a teacher helps him guide it and sometimes something wonderful happens."

Teachers had also helped John Vellinger and Lloyd Bruce, the two other students with experiments flying on *Challenger*. Vellinger, a sophomore at Purdue University, had begun designing his study of eggs when he was a ninth grader in Lafayette, Indiana. He had redesigned it as a high school sophomore and then as a junior, the year it was accepted. Now he was a sophomore at Purdue University. Bruce, a sophomore at the University of Missouri, had waited nearly as long to fly his experiment on strength in metals, but he admitted he was more excited about the possibility that he had come "one step closer to experiencing my dream of becoming an astronaut."

Each of them thanked his teachers.

"It takes a special person to bring out the best in a student," said Cavoli, whose parents, Italian immigrants, ran a furniture shop in upstate New York. "I'm proud my experiment is flying with the first teacher in space. She's helping teachers everywhere finally get the glory they deserve."

The plan for the second day of briefings was to divide the journalists by category — morning newspapers, afternoon papers, national magazines, wire services, television, radio — and place them in separate government offices for a series of round-robin interviews with the individual members of the *Challenger* crew. Each group would spend twenty minutes with each crew member, except for Scobee and Smith, who would make the rounds together.

Breathing heavily and looking as if she had not slept well the night before, Resnik arrived first, wearing a dress with long sleeves, brown-and-white stripes and a high collar, her face pale and drawn. She popped open a can of diet Coke.

Resnik believed, like most astronauts, that she had nothing to gain and everything to lose by baring her soul to the press. The process by which astronauts were assigned to shuttle missions was mysterious. Some of them had been in the corps for years and had yet to fly while others had flown three or four times. None of them knew what made the difference, and they feared that uttering an untimely remark to the press would ground them for eternity. Few of them felt as strongly as Resnik, however. She told friends that interviews made her feel like "a potted plant."

"You're a payload specialist on this flight, aren't you?" a reporter began.

"A Mission Specialist," Resnik said, her jaw tightening.

"What about the malfunctions in the shuttle mission simulator?" another reporter wanted to know. "Did you find them to be difficult, or at least challenging?"

"They're challenging," she said, "but I wouldn't consider them to be difficult. It's just a matter of understanding the systems and practicing a lot. And that's really how we train; we practice a lot."

"But you practice so much that when you actually get there, isn't it like . . . ?"

"It's just like you practiced it," she said, "except usually you practice all the things that can go wrong, and usually nothing goes wrong."

"But don't you have the feeling, 'Heck, we've done this all before?' "

Resnik rolled her eyes. "Yes," she said, "and that's the way it should be."

"What about the ice, though? Did you practice for that?"

Hazardous ice chunks had formed on her first space flight in orbit beneath the wings of the shuttle *Discovery*. Resnik and the commander, Hank Hartsfield, had used the remote manipulator arm to chip the ice away before reentry, perhaps saving their lives.

"We had trained for almost every possibility except the ice," she said. "But part of training is learning what the arm can do for you and where it can reach. It was just like any other task, only the ice was in a different place."

"Do you find it challenging to confront a problem like that?"

"I would rather the mission just went perfectly," she said.

When a reporter asked her how Christa might humanize the space experience better than the astronauts had, Resnik sipped her diet Coke and stared at the floor for a second.

"I really don't know exactly," she said. "Since I haven't talked to her about that, I'm not sure what she's referring to. You ought to ask *her* what she thinks."

"Well, since she hasn't been up there, she may not know. What do you think?"

"I think the astronauts who have flown have come back and have talked about what they have done about as well as anybody could have. We're technical people. Our careers are being astronauts and engineers and scientists, and once we have talked about it and told everybody what we did on a particular mission, then it's time for us to move on and do something else. Part of Christa's job is to talk about it. Most of our job is not to talk about it. It's to do it."

"What about being a woman astronaut?" she was asked. "Have you talked to Christa at all about it? About dealing with the media, maybe?"

"Well, as far as the media is concerned, I think she's dealt with it right from the start. I don't think she needs any advice from me. And as far as the other question, I'm just a person doing my job."

Period.

A veteran space writer reminded Resnik how reluctant she had been to meet the press before her first mission. He asked her if she was more relaxed now. She pursed her lips.

"I've always felt fine about it," she said.

Then she left without saying good-bye. One inquisition had ended. She had five more to go. It was not a good day for Judy Resnik.

Scobee and Smith blew in like a fresh breeze a moment later.

"Did someone get your descriptions mixed

up here?" a reporter deadpanned, waving their capsule biographies in front of them. "They're identical: brown hair, blue eyes, six feet one . . . "

"Don't you know all pilots look alike?" Scobee said, laughing.

"They all weigh the same, too," added Smith.

The tension of the Resnik interview had lifted, and someone asked Scobee what he enjoyed most about space flight.

"I guess it's just getting to do something a lot of other people don't get to do," he said. "There's the excitement of the launch, because there's all that disassociated machinery, hopefully all going off at the same time . . . "

"And all built by the lowest bidder," a reporter reminded him.

"Yes," he said, smiling. "And all built by the lowest bidder."

One thing had bothered him about his first shuttle mission, though — coming home.

"It's really a nice, exciting time, and you return with a lot of memories, but one of my consternations about coming back is that right afterward you have to go tell everybody about it. Sometimes I feel like I'd just like to lock it up inside and keep it for myself and not keep telling it, because every time I tell it, it gets a little bit older hat and it takes some of the fun out of it."

Someone wondered how Mike Smith felt waiting five years to see how much fun it was.

"Well, it's been a little bit long," he said, "but we were told it was gonna be five years when we got here, so what can I say?"

"Are you excited?"

"Sure. You bet."

"Do you have kids?"

"Three kids."

"Are they excited?"

"Sure," he said. "They're really anxious to get to Florida so they can go to Disney World."

When everyone had stopped laughing, Scobee mentioned that Smith had been chosen to fly *Challenger* again nine months after mission 51-L.

"Will you be the commander?" Smith was asked.

"No, I'll be the pilot again, but that's all right with me," he said. "If they want to fly me every nine months" — his voice suddenly rising — "that's just great."

A moment later, the door opened. Out went Scobee and Smith, and in came Christa. She wore her lucky yellow jacket, her arms swinging in her bouncy stride, her hair bobbing on her shoulders. She was happy to see everyone, but she regretted that it was her last round of official interviews before the flight.

"I have mixed feelings," she said. "I'll miss the interviews, because you people allow me to share what I'm doing with my kids and my students. One of the hardest things for a teacher to do is to experience something and not be able to tell anybody about it. I'll miss that, but on the other hand, everything's going to go real fast from here."

"Are you excited yet?"

"When I started this, I didn't think I could get any more excited," she said, "but it just keeps building and building. I'll probably be a wreck by the time the shuttle goes off."

She was particularly excited about watching the crew deploy the TDRS ten hours after lift-off.

"I just hope there's room on the flight deck for me to see it," she said. "I'm so afraid I'm going to be down on the middeck and they're going to be saying, 'Stay out of the way.'"

Then Julie Morris, a reporter for *USA Today*, wanted to know the status of Christa's journal. She wanted to know when people could read it and where — a book, a magazine, a newspaper or some other medium. Christa balked. The journal had become a sore point. It had been one of the reasons why she was selected for the shuttle ride, a poignant reminder of the American pioneer women who had passed on their personal visions of a new frontier to future generations. People across the country were eager to see space through the eyes of a small-city schoolteacher, and NASA had announced on the day of her selection that the journal would be public property for the year she worked for the agency. Her personal thoughts on everything from the selection process to the flight and the fame would be an open book. That was the idea anyway.

But Christa had other ideas, none of which she had anticipated before her selection. First, she realized her writing skills were weak, and she began to strain under the pressure of recording her thoughts for millions of potential critics. Then as her free time diminished, she chose to write letters to friends instead of writing journal entries. In fact, said her NASA publicist, Linda Long, there was no journal. There were notes — scraps of paper she had shoved in her pockets, strewn about her apartment and left in her desk — but nothing to give the people who were eager to publish her early entries. Requests had already come from *Life*, the *New York Times*,

USA Today, the *Concord Monitor* and others.

There was also the privacy issue. Whenever Christa collected the journals she had asked her students to keep, she had told them to fold back parts of the pages that included personal thoughts they would rather not share. She chose to do the same thing in her space journal, a conviction that grew stronger as she adopted the astronauts' code of honor: astronauts did not tell stories about each other for public consumption. If someone suffered space sickness or cabin fever or spilled his chocolate milk all over the orbiter, it was his business, no one else's. Christa planned to protect the crew's privacy as well as her own, and she began to close the book on an idea that had helped her win a ticket to space.

Long warned her of the backlash a decision to not release the diary might trigger, but Christa held her ground.

"Nobody told me they were going to take my journal and publish it," she protested. "I'm just not going to give them word for word what I write."

She intended to use a Dictaphone to record her observations on the *Challenger* flight, but she grew so wary of NASA's intentions that she wondered whether the agency would confiscate the tape when she landed and release it without her permission. Long finally struck a deal with her.

"We agreed she'd release only the golly, gee-whiz stuff after the flight," she said. "It wasn't easy, though. The journal issue was the closest we ever came to an argument."

Then it was Morris's turn.

"Is the journal public or not?" she asked. "Is NASA going to distribute it?"

"From what I understand," Christa said, "it's going to be my own, and I can do with it as I wish when I come back."

"Although at one point you said you understood it was public and belonged to NASA . . ."

"No, they've never told me that."

"So it's not going to be released, as far as you know?"

"As far as I know, no. I'm not going to do anything with it until September 1 [when her contract expired], and even then I don't know how I'm going to use it. I'm not doing it for a lot of people. It's something I need to do for me. I really want to give the personal view to my family. I want my kids to know what all of this has been like. Then, as a historian, I'd like to find a way to share the rest of it with the public. I've thought about maybe using it as a curriculum guide for a futures course that would reach a wide segment of people."

Then she was gone. Jarvis stopped by to share his excitement about the mission, but he became so consumed by explaining the fluid dynamics experiment he would conduct on the shuttle that the interview ended before his explanation. Next came McNair and the only public controversy between members of the *Challenger* crew.

McNair had taken aboard his first space flight a small soprano saxophone and recorded three songs while he orbited the Earth. He had planned to share them with his family and friends when he returned, but he had accidentally erased them on the last day of the mission and had too little time to tape the songs again. He wanted to play them again on *Challenger*, but the saxophone had not been cleared for flight.

"Why?"

"Well, let's say there's some objection," he said. "Someone in the chain of command objects to it this time."

"Too frivolous for a space mission?" said Gayle Golden, a science writer for the *Dallas Morning News*.

"No, it's not frivolous at all," McNair said, his voice rising. "As a matter of fact, it's everything but frivolous. It's something meaningful."

He had planned to play the same songs he had played on his first mission, "America the Beautiful," "What the World Needs Now Is Love" and "Reach Out and Touch Somebody's Hand."

When reporters pressed him on who had objected to the idea, McNair balked. He wanted to avoid friction, he said. He didn't want "to get anybody ticked off." He suggested the reporters use their investigative skills to track the story. They intended to do that, but first they had a few other questions.

"You know, Christa has said nothing worries her yet about the flight, but that the fear might hit her when she walks across the catwalk to get on board," one of them said. "What could you tell her about it that might put her at ease?"

"I could tell her she's right," McNair said. "She'll walk out there and see those big boosters hanging there and the big tank and all that huge stuff smoking, and she's gonna know it's for real. At that moment, you're hit with a heightened awareness of how alone you are."

"You never hear astronauts talking about being afraid," Morris said.

"Well, I guess you don't hear about it because . . . " Then he caught himself. The code of honor had suddenly risen before him.

"Well, it's not a frightening thing," he continued. "It's more like fun, a joyride. It's not frightening. The only thing you really fear is a problem that causes the clock to stop counting, because then you have to go back and try again. You want the clock to keep moving. You don't want any problems, because you want to get off the ground. You know, you're not gone until you're gone."

"Do you think it's a good idea to put ordinary citizens on space shuttles?"

"Sure. Why not? Let everybody have a little bit of the fun."

"Do you think the space program needs to be explained in the words of ordinary citizens?"

"Well, I think there are lots of people to explain it, but Christa will get a lot of other people to listen, and therein lies the key. Many people, like the educational establishment, who were perhaps apathetic or uninterested before, will be tuned in this time. They're going to be informed whether they like it or not, just by watching. There'll be a lot of, 'Oh, I didn't know that.' A lot of them still think we're in capsules eating out of toothpaste tubes."

Then the final question: "Is there a space culture?"

"Space culture," McNair said, twisting up his face. "No, I don't think so. At least I haven't discovered it." He said good-bye and hopped out of his seat.

"Space culture?" he said to himself as he headed for the door. "What an interesting question."

Now he was walking alone in the hall.

"Space culture?" he said again. "No, I don't think so."

During a break the reporters found Scobee and Smith drinking coffee from Styrofoam cups and leaning against a corridor wall. Scobee pushed himself off the wall as they approached.

"Excuse me, Dick," one of them said, "but I was wondering if you could help me unravel the mystery of Ron McNair's saxophone."

"It's no mystery," he said without a trace of animosity. "It was my decision. We just don't have the room for it. We're packed tight enough as it is."

Scobee sipped his coffee, smiled and added, "Let's put it this way. I decided Ron could bring his sax if Judy could bring her piano."

Someone else asked him how he felt about payload specialists and space flight participants.

"The idea really torqued me at first," he said, "because we had people [in the astronaut corps who] had been waiting fifteen years to fly and lost their seat to somebody who walked in off the street. It was a shame, but now most people have flown, so the stigma has kind of gone away. All I ask is don't fly me with a politician."

Onizuka explained in the final interview of the day that flying a teacher was as dangerous as it was historic.

"One should never interpret space flight as routine," he said. "Every mission has its hazards. It requires a lot of training and hundreds of people working together on the ground to make sure it's not a catastrophe. There are still a lot of questions about the shuttle, and we need to make sure it's as safe as possible before we start flying too many civilians."

Still, he said, he wouldn't mind a malfunction or two once *Challenger* reached orbit. He had spent hours underwater training for an emergency space walk and was anxious to try one.

"You mean you hope something goes wrong?"

"Yes, being an EVA [extravehicular activity] type, yes. I'd love to go out there and shimmy up that [remote manipulator] arm.

It'd be just like climbing a coconut tree."

The treetops had given Onizuka a better view of Hawaii when he was a child. Now he imagined a space walk opening a new window on the universe. He itched to see what he saw on his first flight, only better.

"I saw things I never dreamed I'd see," he said. "I saw some of the most beautiful sunrises and sunsets you can imagine. I'll remember those pictures forever. God, will I remember them."

The most outgoing of Christa's crew mates, Onizuka enjoyed bantering with reporters, but he was as eager as Scobee was to press on. When the interviews ended, the *Challenger* seven continued their flight preparations, some as complex as mock disasters in the shuttle mission simulator, others as routine as deciding on the few personal items they were allowed to take into space. Each crew member was limited to twenty items that all together weighed less than a pound and a half and fit into a small storage tray. Christa ran into a problem — Fleegle.

Her personal flight kit was too small for Scott's stuffed toy frog. Scott wanted dearly for Fleegle to fly, and when Christa sensed his disappointment she promised to find a way to take the frog along. She liked the little animal, too. She wanted to take a picture of the frog floating in space.

"He's green and fuzzy with his tummy hanging out," she explained. "And he has such cute little puffy eyes."

Scott gave her a backup frog, Fred, in case Fleegle was grounded, but after lengthy negotiations with NASA and her son, Christa agreed to take Fleegle on the condition that his insides were removed and he was vacuum packed to fit into the flight kit.

"Scott is *so* excited," Christa announced. "Fleegle's his prize possession."

Even vacuum packed Fleegle filled up much of the tray, but Christa managed to find him plenty of company. Steve gave her his class ring from the Virginia Military Institute, and Caroline gave Christa the cross and chain she had received at birth. She took a ring for her sister Lisa and a Girl Scout membership pin for the daughter of her friend Anne Malavich. Steve's brother, Wayne, a Navy pilot, gave her his aviator wings, and Steve's brother-in-law, David Winfrey, sent the unit pin he wore as an Army physician's assistant. For friends and relatives like her father, who wanted her to take his wedding ring but was unable to get it off his finger, Christa added silver medallions she planned to give them when she returned.

For herself, Christa packed the watch her grandmother had given her and a copy of the poem "High Flight," written by a Canadian combat pilot who had died over Britain in 1941. She liked what it said about exploring the heavens:

Oh, I have slipped the surly bonds of Earth,
And danced the skies on laughter-silvered
  wings;
Sunward I've climbed and joined the tumbling
  mirth
Of sun-split clouds — and done a hundred
  things
You have not dreamed of — wheeled and
  soared and swung
High in the sunlit silence. Hov'ring there,
I've chased the shouting wind along and flung
My eager craft through footless halls of air.
Up, up the long, delirious, burning blue
I've topped the wind-swept heights with easy
  grace,
Where never lark, or even eagle, flew;
And while with silent, lifting mind I've trod
The high, untrespassed sanctity of space,
Put out my hand, and touched the face of God.

Christa squeezed several items — miniature

flags for Concord High School, Marian High School, Framingham State College, the Council of Chief State School Officers, the city of Concord and the State of New Hampshire; pins for the Concord Education Association and the New Hampshire chapter of the National Education Association; patches for the Concord Police Department and the New Hampshire National Guard, in which Steve served; and the official seal of the town of Framingham — into the crew's official flight kit when she ran out of room in her personal souvenir tray. The kit also included fifty miniature American flags, forty-seven copies of the Constitution, two sets of 1986 Liberty coins, flags from each of the other crew members' hometowns and states, the key used to dedicate their launch pad and a deflated soccer ball for the Clear Lake High School girls' soccer team. Mike Smith's daughter, Alison, was a member of the team.

To help her relax at her orbital bedtime, Christa brought a portable cassette player, headphones and tapes of three of her favorite musicians: Carly Simon, Bob Dylan and the Irish flutist James Galway. She declined Steve's request to take a tape by one of his favorites, Leon Redbone, a traditional jazz and blues singer.

NASA also allowed her to add two T-shirts to the wardrobe she would wear aboard *Challenger*. One of them was emblazoned with the New Hampshire State seal. The other said, "I Touch the Future — I Teach."

It was mid-December by the time Christa had finished packing for space flight, and she needed to catch up on her Christmas shopping. It hardly seemed like the holiday season, though. After a lifetime of New England Christmases, something bothered her about shopping in 80-degree heat and seeing children sit on the lap of a Santa Claus who wore shorts and a short-sleeved jacket. What next, she thought, a beardless Saint Nick?

Still, she found a few things she liked for the kids: a tiny mechanical pig that would do tricks for Caroline, and a couple of frogs for Scott's collection, including one that was dressed up like the frog prince of Grimm's fairy tale. For her nieces and nephews, Christa raided the space center gift shop's supply of T-shirts, model shuttles and astronaut ice cream. She bought "The Dream Is Alive" calendars for about two dozen friends and relatives, and before she wrapped them she placed a teacher's gold star on January 22, the day she was scheduled to leave for space.

On December 19, though, came the first

hint that the date would change. A countdown to the launch of the shuttle *Columbia* stopped suddenly that morning when a faulty steering unit in a solid rocket booster forced NASA to abort the mission fifteen seconds before lift-off. The flight was rescheduled for January 4 to give shuttle workers time off for the holidays. Christa's mission would not be delayed, NASA officials announced, because *Challenger* was to take off from a newly refurbished launch pad a mile from *Columbia*'s. Their optimism soon proved groundless.

Back in Concord, meanwhile, Steve was less concerned about launch delays than he was about buying a Christmas tree before Christmas. The days had rushed by as he strained under the pressures of a busy law practice, the holiday season and single parenthood, and now he could wait no longer. He rushed after work to pick up the kids and head for the tree lot. The temperature had plunged to near zero, so he bundled them in several layers of clothes, squeezed them into the car and turned up the heat. In his hurry, he forgot to turn it down. Soon Caroline was sweltering, and the combination of the heat, the fast pace and the excitement upset her. She began throwing up at the tree lot, Steve comforting her with one arm as he chose a tree and a wreath with the other.

"I'll take that one and that one," he told the lot attendant, pointing blindly over his shoulder in "the fastest Christmas tree sale on record."

Christa came home several days later ready to rescue Steve and settle another matter — a dispute that had erupted over the city's plans to welcome her home from the *Challenger* flight. A group of enthusiastic city officials and businessmen, eager for the international media to trumpet the glories of Concord as well as Christa, had formed a committee to orchestrate a homecoming for her on the scale of the 1984 Olympic Games in Los Angeles.

"I see it as a golden opportunity for the city to put its best foot forward," said Van McLeod, a theatrical producer and a subcommittee chairman.

The plans featured a parade, a speech to the legislature, fireworks, entertainment and fifteen thousand copies of an eight-page full-color program. Between events, members of the press would be treated to a banquet and a bus tour of the city. The celebration would require six full-time workers for several months and cost between $80,000 and $173,000, which the committee planned to raise through individual and corporate contributions, about $26,000 in city money and the sale of a copyrighted REACH FOR THE STARS logo. A suggestion that money be raised from merchants by promising them that Christa would appear at their stores had been nixed by Linda Long, who said the idea "gave me the hives."

The plans troubled Christa, too, and others in Concord as well. The *Concord Monitor* said in an editorial titled "Hype-Hype Hooray" that businesses were free to try to "make a buck off the McAuliffe flight," but that taxpayers should be spared such entrepreneurial fanfare.

"We wonder if the city isn't sacrificing genuine feeling to promotional hype in an effort to use McAuliffe to 'sell Concord,'" the editorial said.

An emotional debate erupted among city leaders, and through her lawyer, Leo Lind, Christa told the committee that the celebration was headed in a direction she "hoped to prevent." The committee withdrew its request for city money, but hard feelings lingered. Now Christa was home, and NASA was about to break tradition again to let her address the issue, shelve it and focus attention back onto her mission.

"This is a special occasion to do this within

thirty days of a launch," said Ed Campion, welcoming three dozen reporters to a press conference at the Ramada Inn. "But Christa is a special person."

Sgt. John Clark checked under the podium for a bomb before Christa strode in with police officers at her side and stood before a bouquet of microphones, a huge NASA banner hanging on the wall behind her. Her students — a half moon of journalists — stood before her while Caroline, wearing a white party dress and bobby socks, sat on a chair behind them with her babysitter. Christa said she would appreciate as much coverage of her lessons as possible. She described some of her shuttle activities and explained how her students at Concord High would be able to ask her questions by satellite after her lessons.

"They'll be able to pick up a telephone and say, 'Mrs. McAuliffe, how do you feel?' " she said.

Then, in a motherly voice, Christa said, "Oka-a-a-y? That's about the end of what I wanted to talk about, so I'll open it up to questions."

And she was ready for them. Christa no longer suffered the insecurity that prompted her to practice before Roger Jobin's video camera. She no longer walked away from television interviews with microphones attached to her blouse. She had the same boundless enthusiasm, the same folksy humor, but now she fielded questions like a pro. The first ordinary citizen bound for space made it clear she wanted no extraordinary homecoming.

"I'm hoping it's a very simple celebration, because that's what Concord's all about," she said. "This isn't New York City. It's not the Olympics. It's Concord, New Hampshire, and a homecoming should reflect the community I'm part of. I was thrilled when I came back in July and it just happened to coincide with the parade. That was nice. I was able to see a lot of people."

It troubled her that anyone would spend $173,000 for such a party while teachers worked two jobs to make ends meet.

"You don't really need money for a celebration," she said. "It doesn't have to go hand in hand. Teachers are often given monumental tasks with little or no money, and we're able to do a fantastic job. I'm hoping this goes off as simply as possible with as little money as possible."

"Does that mean city money?"

"Any money," she said.

A teacher to the end, Christa then answered a question about the value of the space program by asking everyone to look at the Velcro on their carry bags, and when someone asked her about cutting NASA funding, she conducted a history lesson.

"There's no way you can stop people from exploring," she said. "There's no way you can stop that excitement. Think about the early explorers who went to find a queen or a king or somebody to subsidize them. If NASA didn't do this, we'd have businesses doing it. People are always going to explore."

And when a television reporter asked her in all sincerity if she was concerned about the Neilsen ratings of her lessons from space, Christa fixed him with a look she reserved for classroom pranksters.

"No," she told him. "Not at all."

A few minutes later, she said she was happy to be home, "where it's pretty sane and normal . . . and where you know you can come back and be accepted as part of the community." Then she smiled, thanked everyone for coming and followed her police escort out the door, Caroline holding her hand.

The final days before Christmas were crazy, but Christa insisted on spending as many as she could with the kids. On December 21, she took them to see *A Christmas Carol* at the

Capitol Theatre, where at intermission she met Audra Beauvais, her former student and babysitter. Beauvais thanked Christa for writing a recommendation for her and told her she had been accepted to Lesley College. Beauvais told Christa she planned to become a teacher. Christa hugged her and told her she was scheduled to speak at Lesley after her flight.

"I'll see you there," Christa said.

On December 22, she and Steve took the kids to see a New England Patriots game. They had been invited to sit in the owner's box, an offer that for Scott's sake they could not refuse. The temperature was in the teens and an occasional snow flurry fell, but Scott was a Patriots fan — even if he *did* wear the Washington Redskins jersey Steve had bought for him — and the Patriots were on their way to the Super Bowl, an occurrence as rare as Santa Claus appearing in Concord in shorts and a short-sleeved jacket.

After they had watched the Patriots toy with the Cincinnati Bengals, 34–23, the McAuliffes were allowed near the Patriots' locker room so Scott could ask the players for their autographs. Several of them obliged, and then Scott approached Brian Holloway, an all pro tackle whose size — six foot six, 285 pounds — was as intimidating as his sweat-streaked face and his flared nostrils. Scott was scared. Christa went with him.

"Hey, I know you," Holloway said as they walked toward him. "You're the teacher astronaut."

He had recognized her smile.

"It's an honor to meet you," Holloway said, "a real pleasure."

Several other players recognized Christa, too, and they waited in line for her autograph as Holloway told her about his fear of flying and asked if she was frightened by the shuttle mission. She told him she had great faith in the safety of the space program. She said the flight was the chance of a lifetime. Then as Scott collected autographs of the players waiting in line, Holloway asked Christa for hers.

"To Brian," she wrote. "Reach for the stars! I'll be there!"

On the same day, *Challenger* made the slowest part of its tenth journey toward space. Mated with its fuel tank and booster rockets atop a flatbed transporter, the three-million pound shuttle assembly was hauled across the Kennedy Space Center from the Vehicle Assembly Building to launch pad 39-B, a 4.25-mile trip that took eight laborious hours. There *Challenger* sat for the next thirty-seven days, a space-age steeple on the Florida skyline.

Christa visited her hairdresser the next morning. Her hair had presented problems in the weightlessness of the KC-135. The long curls had floated into her face and distracted her when she turned her head quickly. She needed a tight perm, she told the hairdresser, Steve Gelinas, something strong enough to withstand the anarchy of anti-gravity. Christa explained it was her last haircut before the flight.

"She was all pins and needles in excitement," Gelinas said. "She couldn't wait to go."

As he cut her hair, she addressed a stack of last-minute Christmas cards, and when a customer congratulated her on being selected for the *Challenger* mission, Christa smiled and said, "It doesn't seem possible, does it? A schoolteacher from Podunk, U.S.A., going into space?"

Gelinas wished her luck when she left. Christa told him she would see him when she got back and warned him playfully that if her hair bothered her on the flight, "I'm going to blame it on you from space."

"I hope it does," he told her. "I'd love to hear my name on the radio."

She had presents still to buy, so on Christmas Eve she left the kids with friends and rushed down the hill to a side street where in one short block she could browse through a toy shop, a candy store, a jewelry store, a dress shop and a bookstore. On her way out of the bookstore, Christa met Matt Mead, a former student who had run through his neighborhood and shouted for joy when she was selected. Matt remembered that whenever Christa had left the room in the middle of teaching a law class to retrieve something she had forgotten in the social studies office down the hall, she had continued her lesson as she walked, her voice growing fainter as she moved farther from the roomful of puzzled students. And he admired the way she wove her personal experience into her lessons. He remembered a discussion on the legalization of marijuana when Christa had explained that her father had legally smoked it several years earlier to combat the nausea he suffered from chemotherapy he had undergone for lymphoma, a form of cancer he had conquered through treatment. To the students' delight, Christa had described her father's pulling down all the shades in the house to keep his neighbors from seeing him puff the pot.

Matt had sent Christa a congratulatory card on her selection, and he had attended the celebration in her honor at the state house plaza. In mid-December, he had sent her a letter in Houston and asked her to write a college recommendation for him. She had not yet returned it and now apologized to him when she met him on the street, her arms full of bundles. Christa told him it was on her desk in Houston and that she would mail it to him after the holidays. Then she wished him a merry Christmas and told him she would see him at the high school graduation in June.

Across the street she met Neil Harmon, one of her former English students at the junior high school in Bow. Harmon had learned the value of keeping a journal from Christa, and later he had parked his bicycle in her garage when he attended Concord High. He had gone away to college to study engineering but had recently decided on a new career. He was eager to share the news with Christa.

"How are things going?" she asked.

"You'll never guess what field I'm going into," he said.

He was right; she had no idea.

"Teaching," he said, smiling. "I'm going to be a teacher."

He had told several other teachers about his decision, and they had warned him about low morale and low salaries. But Christa beamed.

She congratulated him and wished him well.

"He sounded so excited and proud," she said later. "And I felt great."

It was the night before Christmas and Christa had renewed a four-year tradition by inviting John and Marcia Rexford, neighbors with no children and no relatives in the area, to dinner. As she prepared roast beef in the kitchen, Christa heard voices outside. It was too early for the Rexfords. She wiped the frost off a living-room window and saw that a group of her former students had gathered on the sidewalk to sing Christmas carols. It was cold and dark, but Christa tramped through the snow without a coat to join them, hugging each of them as she sang along, smiling and shivering. She lingered for a while to ask them how they were. Then she passed out Christmas candy and walked back to her doorstep.

"Merry Christmas!" they shouted.

"Merry Christmas," she said, silhouetted by the soft light of her living room.

The Rexfords arrived at seven o'clock, and when Christa opened the door, she heard the bells toll from a distant chapel. The house was rich with the scent of pine. The tree lights twinkled, and Marcia Rexford remembered a Christmas several years earlier when Scott had ridden about in his little red fire truck. Marcia hadn't expected to visit the McAuliffes this Christmas. She hadn't seen them since early November, when Steve and Christa had run out of gas in front of their house and stopped in for help.

"Don't forget about Christmas Eve," Christa had said as she hurried off to catch a plane back to Houston.

"Christa, you have no time for that this year," Marcia said.

"Well, I'll give you a call," Christa said, and she did — two days before Christmas.

When the McAuliffes had stopped at the Rexfords in November, Steve had admired a picture of John, which had been taken by a local portrait photographer. Marcia had called Steve later at work and asked if she could take the photographer to the babysitter's house and have a portrait made of Scott and Caroline as a Christmas present for them. She gave it to them on Christmas Eve, and Christa was fascinated by it. She dashed upstairs and returned with a picture of her mother's father. Her mother's parents, who were of Syrian ancestry, had died young, leaving her to be raised by her grandparents. Christa had never met them, but now she saw something in Scott's eyes that were strikingly similar to her grandfather's. They were dark brown, almost Oriental. They were sensitive eyes. How wonderful, Christa thought, that the fabric of the family had endured the march of time.

It was a night for nostalgia, and when Marcia gave the children handkerchiefs filled with little trinkets, Christa recalled her grandmother's taking her up to her room each time she visited and letting her choose a new handkerchief out of a drawer. She was sorry her grandmother could not see her now; there was so much she wanted to tell her.

Christa read a children's book to Scott and Caroline after they had feasted on roast beef and microwaved potatoes and mincemeat pie. It was Christmas Eve in the story and Santa Claus was preparing to make his rounds on the space shuttle.

The next morning, they were too late for Christmas mass at St. Peter's, their parish church, so the McAuliffes rushed to St. John's, a mile away. With 350 others, they sang "Silent Night" and "Joy to the World," and repeated the psalm response "All the ends of the Earth have seen the power of God." Someone whispered during communion to the Reverend Daniel Messier that the space teacher and her family were in the congregation, and

Messier, who had seen Christa only in pictures, scanned the pews in search of her. He spotted her quickly.

"It was that unmistakable aura of excitement," he said. "You could tell there was real ecstasy inside of her."

Afterward, he joined Christa, Steve and the kids on their hands and knees, crawling between the pews to search for an earring Caroline had lost. The McAuliffes were the last ones to leave. They spoke with well-wishers on the sidewalk, and then drove an hour and a half to Framingham for a quiet Christmas with their families. Everyone laughed at Christa's space-related gifts, and she showed no disappointment at not receiving the lamps she had wanted. Steve gave her something better: jewelry. She had complained to friends that he rarely gave her jewelry, but this year he surprised her with a set of gold earrings. They were shaped like apples and she loved them. She took them with her on the shuttle.

The next week was Christa's last free time before the launch, so she spent most of it alone with her family. She baked cookies with Caroline, watched Scott's hockey practice and drove back to Massachusetts to help her five Girl Scout friends throw a surprise party for her mother, who had brought them together thirty years earlier. On the afternoon before New Year's, Christa visited Ginny Timmons, her next-door neighbor, and drank tea with several women from the neighborhood. They talked about their children, their homes and cross-country skiing. There was little mention of the space program, but Christa autographed a picture of herself for Timmons's fourteen-year-old daughter.

"To Jeanne," she wrote. "May your future be limited only by your dreams. Love, Christa."

Then Christa rushed off to take part in the city's First Night celebration. The year before, when Steve was nagging her to complete her teacher-in-space application, she had marched down Main Street in a costume parade as the spider from *Charlotte's Web*. This year Christa had agreed to present the trophy to the winners of the annual snow sculpture competition, whose theme was "Reach for the Stars." Workers from four companies had spent several days creating the sculptures on the state house lawn. There was a giant shuttle with a cockpit in which children could play and pose for pictures. There was a shuttle lifting off from the Earth through a sea of suspended stars. And there was the winner, "The

Starslide," which allowed children to "reach for the stars" by climbing a ten-foot mountain and to "experience the excitement of a space adventure" by entering a cave on the summit and returning to Earth on a steep, ice-slickened slide. Caroline had personally approved it.

Meanwhile, Steve had spent the day — yes, the day — cooking spaghetti sauce for a small dinner party with Leo and Margaret Lind and the kids. Out of habit Margaret Lind had brought some of her own sauce and Steve decided to put himself to the test. He would ask Caroline, a spaghetti lover, to judge which one of the sauces was better. Spaghetti sauce had become his specialty, and this was a chance to show Christa just how much the kids had enjoyed their father's cooking while she had been gone.

Caroline had no intention of hurting her father's feelings, so when she sampled the sauce she believed was his, she rubbed her stomach and gushed, "Oh, it's great."

It was Lind's, but no one had the heart to tell Caroline. While Steve tried to stifle his embarrassment and everyone else smiled, Caroline took another spoonful.

"Oh, Daddy," she said, "you're the best in the whole wide world."

The next night, six months after Christa paraded down Main Street on a summer day of sunshine and promise, she left home for the last time. It was cold and gray, the kind of day when you wondered if summer would ever come again.

CHAPTER NINE

By all accounts, 1986 was to be the year of the space shuttle. Two new launch pads — one at the Kennedy Space Center, the other at Vandenberg Air Force Base in California — would help shoot the shuttle toward the stars fifteen times in twelve months, a runaway record for manned space flight. Scientists would celebrate the orbital debut of a space telescope that could see fourteen trillion miles and advance astronomy further than any instrument since Galileo's first telescope in 1609. The president would proclaim the world a safer place after four military missions enhanced America's "Star Wars" strategic defense system. People everywhere would herald the flight of a schoolteacher who humanized the high frontier. NASA would reclaim its faded glory.

Then came *Columbia,* which started the year by setting a space agency record for futility. Seven times in twenty-five days the star-crossed orbiter was scheduled to enter space, and seven times it failed to get off the ground. With only four shuttles and a lengthy turnaround time between missions, NASA's bold promise for the new year already seemed beyond its power. But red-faced officials vowed to press on as journalists, once loyal boosters, razzed them. Newspaper columnist Art Buchwald noted that shuttle flights had become as dependable as the notoriously undependable Long Island Railroad, and the network anchormen were no more kind.

"It's now zero for five for the shuttle *Columbia*," NBC's Tom Brokaw announced after the fifth delay.

"The launch has been postponed so often," said Dan Rather of CBS, "that it's now known as *Mission Impossible*."

And ABC's Peter Jennings put it in historical perspective: "The space agency hasn't had such problems since the days of the Mercury program."

Through the cacophony of criticism, however, rose an encouraging voice. The delays gave Steve McAuliffe strength.

"I have found them very comforting," he said. "It seems to me they demonstrate beyond anyone's normal patience that NASA is not sending people up unless everything is perfect."

Christa knew the delays were vital to safety, but her patience, never her strongest trait, soon wore thin. In a ritual as stressful as planning her wedding, she had invited 350 friends and relatives to witness her launch. They had reserved hotel rooms and plane tickets, and many of them had borrowed money to make the trip. A series of postponements would push them further into debt. Worse, some of them might miss the launch.

By early January, after *Challenger*'s scheduled lift-off had slipped from January 22 to 23, to 24 and then to Saturday, January 25, Christa's frustration surfaced. She had stood in the predawn darkness on the roof of Kennedy's Launch Control Center on January 6 to see *Columbia*'s third attempt at space flight fizzle. The mission was scrubbed thirty-one seconds before the scheduled launch when controllers discovered that the spaceship had too little fuel to reach orbit.

"I was hoping, probably selfishly, it would go," she said, and speculated that her own launch would be delayed until February, maybe even March.

Her parents had driven south just after New Year's and settled into a condominium at Cocoa Beach, a few miles from the space center. The delays would trouble them less than they would her brother Steve, who was due back in California for his law boards on January 28, and everyone else, particularly teachers from across the country, who lacked the luxury of open-ended vacations.

Christa worried, too, about Scott and his classmates. An enterprising parent had wrangled an all-expenses-paid field trip for the third graders, a front-row seat to the historic mission. United Airlines offered free rides for the nineteen students and thirteen chaperones. A New Hampshire bank, Numerica Corporation, agreed to pay for their food and lodging. Puma, Nike, Adidas, J. C. Penney and Sears supplied clothing and footwear. The Amelia Earhart Luggage Company provided identical sets of suitcases. Dozens of businesses donated more than twenty thousand dollars in goods and services, among them the nonprofit Young Astronaut Council, which chartered the class as one of its chapters.

What if the children waited for a flight that never flew? The idea was to excite them about the space program, Christa thought, not to introduce them to technology's imperfections. *Columbia*'s fits and starts were not a good omen.

Her crew mates had invited nearly three

thousand people to the launch, including more than sixty of Onizuka's friends and relatives from Hawaii. None of them was pleased by the delays, but the astronauts, accustomed to such inconveniences, accepted them as part of the job. Besides, Scobee said, the delays did offer one blessing: more time to train.

"Now we can practice till we get it perfect," he said.

On January 8, four days before *Columbia* finally lifted off, the *Challenger* seven began their last major training exercise — a dress rehearsal of the launch. It began in Spartan dormitory rooms on the third floor of Kennedy's Operations and Checkout Building, rooms where space-bound crews since the Apollo days had slept on the eve of a mission. Rising before the sun, they ate a light breakfast, climbed into fire-resistant flight suits and rode an elevator to the first floor, smiling for photographers as they walked out a cement ramp to a waiting silver van.

Challenger, poised on its tail, waited six miles away, ready to christen a refurbished launch pad that had not been used since 1975, when Apollo astronauts lifted off for an orbital rendezvous with Soviet cosmonauts. The Atlantic splashed against the sand a stone's throw to the east. The mouth of the Banana River lay just to the south, Mosquito Lagoon just to the north. Buzzards and brown pelicans fluttered above the chunk of concrete, where citrus trees had been planted by pioneers centuries earlier. The crew stepped from the van to the launch tower's elevator.

They crossed a steel catwalk 150 feet up to a sterile white room where technicians strapped them into safety harnesses, issued them skullcaps and helmets and practiced saying goodbye. To Christa's delight, one of them gave her an apple; to her dismay, she sat for two hours on the windowless middeck, flat on her back as the crew simulated the final phase of the countdown. The ordeal prompted her to regret *Columbia*'s delays for another reason.

"Sitting there on your back, even in a dress rehearsal, is hard," she said. "When you figure how many times [the *Columbia* crew] has done it during a real countdown . . . well, I have a lot of empathy for them."

The rehearsal ended with emergency escape training, a space-age circus act in which the *Challenger* crew leaped from the shuttle onto mobile platforms and jumped into baskets attached to two-thousand-foot guy wires. In an emergency on the pad, they would slide down the wires at 55 miles an hour to a waiting armored personnel carrier that would protect them from fire. In training, they slid only a few feet to simulate the drop. Then they rode the elevator and took turns driving the personnel carrier.

Christa was at it again, doing what no ordinary citizen had done, coasting across the concrete in a giant, twelve-wheel tank, her windblown face peeking out from the hatch, a study in concentration and pride, a front-page picture in the next morning's *New York Times*.

"I can't believe it," she said, climbing down from the tank. "I wish all my students had a chance to do that!"

Christa entered quarantine a week later in Houston, not the manic isolation of NASA's early days, but a casual confinement that allowed crew members to come and go almost as they pleased. They were required only to eat meals prepared by NASA's chefs and to stay at least six feet from anyone who was not a medically approved "primary contact," particularly children and crowds, the most common source of germs. Crew members with children slept in trailers at the Johnson Space Center. Christa, childless in Houston, slept at Peachtree Lane, where she studied her flight data file, started packing her belongings and continued her nightly calls to Concord.

When she called home on January 17, eleven days before the *Challenger* launch, she said good night to Scott and Caroline, then chatted for a while with Steve, who told her a reporter from the *Concord Monitor* was there to interview him. She had become a friend of the reporter and so asked to speak to him. Christa was thrilled, she told him, that she had won a beer from Dick Scobee by betting on the Patriots against Scobee's favorite team, the Los Angeles Raiders, in a recent playoff game. She was still puzzled that the Patriots had lined up for her autograph, she said, and she was eager to return to school and sign hall passes instead. She was homesick.

"How about the flight?" the reporter said. "Are you as excited as ever?"

"Oh, I can't wait."

"Have fun," he told her.

"I will," she said, her last words to him.

The Celtics were on television, but Steve interrupted the game before the kids went to bed so he could show them an advance tape of an hour-long national television documentary, "Teacher in Space," produced by WBZ in Boston. Burgess Meredith narrated the show, his raspy voice charged with emotion.

"This is the story of a high school teacher

from New Hampshire," he said, bass drums rolling in the background, "who was chosen to be the first private citizen in history to fly in space."

The show opened with Christa signing autographs before the ninth *Challenger* launch, and it ended with the camera moving down Main Street in Concord on New Year's Eve to the beat of Bruce Springsteen's "My Hometown." But Caroline was too tired to watch and Scott was too concerned with the Celtics, so Steve soon turned it off and tried to get them to bed. Caroline kissed him and went dutifully; Scott balked. He camped on the couch, his red pajamas covering his feet, insisting the game, which had just begun, would soon be over. Finally, Steve hugged him, led him to his room and chased down Rizzo, the family cat. Scott needed Rizzo with him to go to sleep.

Wading through a clutter of Christa memorabilia — newspaper clippings, magazines, NASA materials, videotapes — Steve returned to the family room with a beer to talk to the reporter. Just as he sat down, a forlorn Scott called him.

"Daddy," he said, "the cat won't sleep with me."

The year had not been kind to Scott. He

had enjoyed neither his mother's absence nor the spotlight's relentless glare. When a *Concord Monitor* photographer had taken Scott's picture at school the day before, the photographer had tried to draw a smile from him by handing him Christa's official NASA picture and asking what he thought about his mother going up in the shuttle. Scott had turned his head and started to cry.

Fleegle and Rizzo had helped comfort him since his mother had left, but now Fleegle was gone and Rizzo was acting strange.

"Daddy," Scott said again.

Steve hesitated, hoping Scott would forget about the cat.

"Daddy," he persisted, "can you get him for me?"

After rising from the couch, Steve slouched back down, realizing it was hopeless.

"C'mon, Bunk," he said. "I can't make her sleep with you."

Scott finally drifted off, and the reporter, who was writing a profile of Steve, asked him for his life story.

"Easy," he said. "Born, suffered, died."

Steve needed little prodding, though, to talk long past the Celtics game about his family, his life in Framingham, his love of flying. He described his first glimpse of Christa, his respect for her, the family's dependence on her. He remembered the summers he drove her to Girl Scout camp, and he said the happiest moments of his life were his wedding and Scott's birth. His saddest memory was the day he was ten years old and his dog died.

He said the best thing about Christa's absence was that he had discovered his children.

"I was one of those fathers who always came home after the kids were in bed," he said. "I've really been surprised by how much better I've gotten to know them. I've learned what they like to eat and wear and play. I've learned what's important to them."

He had mailed their schoolwork to Christa, encouraged them at hockey practice and dance class, attended his first parent-teacher conference. He had taken them trick or treating and comforted them when they got sick in the middle of the night. He had mediated their disputes, shared their triumphs and disappointments. He had learned a little about himself as well.

"It's been tough, really hard," Steve said. "I was really frantic for the first few months, but now I'm starting to get pretty comfortable. When it comes down to it, it's good to know you can make it on your own if you have to."

They left for Florida five days later, Scott bound for a cut-rate motel in Cocoa Beach with his classmates, Steve for a villa at Disney World with Caroline. Steve and Caroline took a quick flight from Boston to Orlando, Steve clutching his video camera, Caroline her teddy bear. Scott and his classmates took a longer route.

Bedecked in identical outfits — from their new white Nike sneakers to their oversized Young Astronaut Council baseball caps — the third graders arrived before dawn at the airport in Manchester, New Hampshire. They carried Battlestar Galactica activity books, *Astronomy Today* magazines, inflatable shuttles, coloring books, Cabbage Patch Kids, things to keep them busy or comfort them. They would not see their parents for a week. They wore canvas backpacks that said CHRISTA'S KIDS.

As the jet engines whistled, they stood on the tarmac, read a poem to thank their sponsors and applauded a bank vice-president after he announced an essay contest for children across the state. The theme was "The Sky Is Not the Limit — What Christa McAuliffe's Flight Means to Me."

From Manchester, the children flew to Chicago with their teacher, their chaperones and two reporters, one of whom asked Scott if he was excited about going to Florida.

"Yes, I'm looking forward to Sea World and the launch."

"Not one more than the other?"

"No, both are about even."

"But your mother is taking off in the space shuttle. Isn't that more interesting?"

"Well, I'm looking forward to the launch, sure, but I've never seen a whale. I'd really like to see the killer whale."

What Scott did not want to see was his mother's flight delayed until Sunday, the day the Patriots played the Chicago Bears in the Super Bowl. He knew her flight was a once-in-a-lifetime chance, but he also knew that if history was any indication so were the Patriots playing in a Super Bowl.

Before they landed, the pilot announced that Christa McAuliffe's son was on board, prompting the passengers to burst into applause. But a couple of them grumbled that in the eight hours it took the students to fly from Manchester to Orlando via Chicago, Christa could have circled the Earth five times.

That hardly mattered to Scott's classmate Jessica LeClerc, who was taking her first airplane ride.

"This is amazing," she said. "Scary, too."

Scarier, she imagined, than flying on the *Challenger.*

"Riding the space shuttle — now *that* sounds like fun," she said.

As the children came out of the clouds above Florida, the space center sprawled beneath them, the idle *Challenger* poised on the edge of the ocean. It was Scott's first look at the machine that would carry his mother toward space. Someone called to him to play a game, but he stayed at the window and studied the launch pad, mesmerized.

A crush of television photographers met him at the airport, an unnerving spectacle that was tempered only by a hug from Ronald McDonald decked out in a space helmet and electronic gear.

"Scott's getting tired of people following him around all the time," his classmate Kyle Dyment said. "He just wants to hang around with us kids."

The rest of Christa's followers checked in with less fanfare at motels in Cape Canaveral, Titusville, Cocoa Beach, towns that had sprung up on the pine flats and mangrove swamps around the space port. Her relatives came from New England, New York, New Jersey, California and Hawaii. Her friends and neighbors came from Concord. So did the superintendent of schools, several school board members, her union leaders, the police chief and the governor. Two Concord High students came — Brian Ballard, the editor of the school paper, and Audra Beauvais, the superintendent's daughter. Former teaching colleagues from Maryland, friends from Framingham State College, her Girl Scout camping companions — all of them came. On hand also was the class of 51-L.

NASA announced on Wednesday, January 22, that the *Challenger* mission had been delayed another day, this time to the morning of Super Bowl Sunday, and this time not because of *Columbia,* which had finally completed its journey four days earlier. The culprit was a dust storm that had cast a hazy curtain over a runway in the Sahara Desert. An emergency several minutes into the launch would require the *Challenger* crew to attempt a landing on a two-mile stretch of concrete in Dakar, Senegal. It would help if they could see the runway.

The delays pleased almost no one but the souvenir merchants who celebrated their greatest windfall since the inaugural shuttle mission. Christa's smiling face was everywhere as people sported teacher-in-space trinkets and *Challenger* memorabilia — patches, posters and pencils, coffee mugs and medals, thimbles, T-shirts and tie tacks, pins, pen knives and pajamas, shot glasses, statues, stickers and spoons.

"I just spent a hundred twenty-four dollars," Christa's sister Betsy Corrigan announced on her way out of a souvenir shop in Cocoa Beach. Her brother Christopher wore a jacket that sagged under the weight of Christa buttons. Her father, who had bought dozens of the buttons when he visited the store earlier, had pinned on the biggest one he could find, his blue eyes twinkling with pride.

Customers stood three and four deep at the counters in the Gift Gantry, NASA's official souvenir shop, among them Roberta Sage, a school librarian from Katy, Texas. Sage, who swung a plastic shopping basket filled with Christa memorabilia, had planned her vacation so she could witness the launch. She said Christa was "the closest thing we have to a hero these days."

When a reporter mentioned he was from Christa's hometown, Sage's eyes widened and she raised her hand to her mouth.

"You mean you've met her?" she gasped. "Well, glor-r-r-ry, you just can't know how excitin' it is to meet someone who has spoken to Christa McAuliffe. Don't tell me. I bet she's just as real in person as she is on the television."

At the moment Sage was slipping souvenirs into her shopping basket, Christa and Greg Jarvis were riding two miles in a rented economy car from Peachtree Lane to Ellington Air Force Base for a flight to Florida. They drew a few stares as they rode out NASA Road 1 past the Putt-Putt miniature golf course, the balding, gray-haired engineer and the smiling schoolteacher, both wearing astronaut suits adorned with American flags, NASA logos and *Challenger* crew patches. They looked like office workers on their way to a costume party. NASA could not have found two more ordinary-looking space travelers.

Jarvis parked in the visitor's section at Ellington, then stepped out of the car in a flight suit that failed to reach his ankles. He wore shiny black penny loafers, Christa powder blue jogging sneakers.

"Hey, nice shoes," an airman shouted.

Had no one told them real astronauts wore combat boots? Of course, but neither of them considered themselves astronauts. They were ordinary people who intended to remain ordinary people. So here was Christa, the first person in NASA history to embark on the first leg of a journey to space in sneakers. And there was Barbara Schwartz, her NASA media coordinator, waving good-bye to her from across the parking lot. Schwartz was not a primary contact, but Christa didn't care. Schwartz was her friend.

"Oh, Barbara," she said, rushing across the pavement and gathering her in her arms. "The adrenaline is really flowing now."

Then Christa ignored the ban on interviews and spoke for a moment with a Houston television reporter. She reminded him of what she wanted children to learn from her experience. Ten months earlier, the odds of her flying to Florida for a space adventure had been longer than she had cared to consider. But she had tested the limits of her potential, she said, and she had not been afraid to fail. She wanted children to know that anything was possible and that nothing worthwhile was gained without effort, commitment and the will to face the risks.

"That's what life's all about," she said.

She said good-bye and carried her own luggage to the airstrip. Wearing her gold apple earrings and clutching a yellow rose given her by June Scobee, Christa waved to a handful of well-wishers and flew out of Houston for the last time, sorry to leave her new friends yet thrilled to begin the next leg of her journey home. Waiting in Florida was the largest group of journalists to greet a space-bound shuttle passenger since Sally Ride.

Christa touched down just before dusk in a corporate jet on a closely guarded runway where *Challenger* was to land ten days later. The rest of the crew zipped in on T-38s: Scobee flying with Onizuka, Smith with Resnik, McNair with a pilot from Ellington. They stepped to a microphone one by one, the shuttle *Columbia* sitting behind them atop a Boeing 747 that had delivered it from California two hours earlier to the delight of Scott McAuliffe and his classmates. Reporters and photographers watched from behind yellow ropes. Commander Scobee spoke first.

"It's a real pleasure to participate in something the Cape does better than anyone else in the world," Scobee said, "and that's launching space vehicles."

His crew mates stood in a semicircle behind him, solemn except for Christa, who smiled and waved to familiar reporters and photographers, mouthing their names: "Hi, Keith. Hi, Shawne. Hi, Jim." She spotted Steve in the crowd and waved to him. Steve had not yet taken a physical needed to become a primary contact. He waved back and snapped her picture.

After Scobee had spoken, Smith told reporters he looked forward "to getting on orbit and getting the secret handshake." Then an unusually relaxed Resnik joked that her crew mates might throw her off the flight if "Steve Hawley's affliction rubs off on me." Hawley had flown on her first mission, which had been postponed two months, and on the *Columbia* flight that had set the record for delays. No one needed to be reminded that mission 51-L had already suffered four delays.

Smiling, Resnik introduced Onizuka, who had slept forty-five minutes the night before because he had driven to Dallas to watch his daughter play in a soccer tournament.

"It's going to be a great mission," he announced. "We're ready to go fly, and thanks for being out here today."

He turned, took a few steps and rushed back to the microphone, blushing.

"I forgot the one thing I was supposed to do," he said. "Here's Ron McNair."

McNair had been a member of the first shuttle crew to land on the Kennedy runway two years earlier. He wanted to say only that he intended to be the first astronaut to land there a second time. Then, smiling, he introduced "the person perhaps you came to see."

Rocking on her sneakers, Christa grinned

and leaned toward the microphone.

"Well, I am so excited to be here," she said. "I don't think any teacher has ever been more ready to have two lessons in her life. I've been preparing these since September, and I just hope everybody tunes in to watch the teacher teaching from space."

Afterward, the crew posed for photographers, and Steve leaned over the rope to toss her a plastic bag emblazoned with the New Hampshire sweepstakes logo. Inside were seven blue T-shirts embossed with the New Hampshire State seal. Christa gave one to each of her crew mates, who seemed puzzled at first, then displayed the shirts and smiled for the pleading photographers.

"Care to say a few words?" a reporter asked Steve.

"It's great to be here," he said, snapping the reporter's picture.

A NASA official raised his hand a few seconds later and the ceremony was over. The crew left for a waiting van, several of them holding hands with their wives. Christa leaned toward Steve and said, "See you later?"

"Maybe," he said, "if I pass [the physical]."

He did, and the next day, after Christa and Scobee had practiced landing in a corporate jet NASA had outfitted as a shuttle simulator, Steve joined the crew and their families for a picnic at a beach house overlooking the Atlantic, three miles south of the launch pad. A NASA spokesman described it as "kind of a group therapy session, the crew's last social meal with the members of the families and the flight team." They drank beer and ate barbecued ribs and potato salad, and they could have passed for another group of vacationers were it not for the heavyset man with the crew cut who stood in the shadows, watching. George Abbey was the flight crew operations director, "the astronauts' den mother." He accompanied each crew to the pad on launch day and greeted them when they landed.

"He manages their emotions," the NASA spokesman said. "He psychs them up or calms them down, depending on their needs."

His job was easy with the *Challenger* crew. The only people who needed their emotions managed were the families of the crew, particularly Christa's. They were new to the business of watching a loved one ride into the sky on a rocket ship.

"It's surprising," Steve said a day after the picnic. "The crew is really up, not the least bit tense. You'd think they'd be nervous, but they're not. I'm getting a little nervous, though, probably because it's so close. I've had trouble sleeping the last couple of nights. Actually, I've been really tense."

Christa's parents, who would not see her again before the launch, admitted to a few jitters as well. After the picnic they returned to the press center to accommodate the nine hundred quote-hungry journalists, most of whom had no access to Christa, Steve or Caroline and were tired of chasing Scott's class through Disney World, Sea World and the space center. Some of them had grown so restive, in fact, that they had begun writing feature stories on reporters from Christa's hometown.

ON THE MCAULIFFE BEAT said a headline in the *Washington Post*. TO HERE FROM OBSCURITY.

So at NASA's request the Corrigans agreed to the first preflight press conference by an astronaut's parents in the history of the space program. They sat outdoors in front of the press grandstand, the forty-three-hour countdown ticking on a large digital clock behind them, a pack of reporters before them. It was Friday afternoon, January 24, and the launch was scheduled for Sunday morning. Ed wore a giant Christa button, Grace a *Challenger* necklace and a nervous smile.

"Christa's just bubbling all over," Grace reported.

"She's happy as a clam," said Ed. "It's like she's been training for this all her life."

Grace had a knot in her stomach, however, and Ed felt "a little bit of trepidation." When a reporter asked Ed how he and Christa had said good-bye to each other, he said he had tried not to upset her.

"I tried to be as casual about it as possible," he said. "I felt she was a little more emotional than she usually is, but she looked beautiful. I've never seen her look better."

The next morning's top story in the *Concord Monitor*, the last edition before the scheduled launch, started this way: "As people from Key West to Concord blink the sleep from their eyes tomorrow morning, Christa McAuliffe plans to cross a 15-story catwalk, strap herself into a giant steel bird and wait for 6.5 million pounds of violent power to hurl her into space.

"Is she frightened?"

The story said she was not frightened, not yet anyway. She admitted fear might strike her once she stepped onto the catwalk, but her faith in the safety of the space program was absolute. Christa told reporters in New York that she would feel safer in the shuttle at 1 in 100,000; an independent federal review panel placed the odds at 1 in 1,000. An Air Force study said they could be as low as 1 in 35. Christa knew only that she faced the possibility of having "not a good day."

Neither did she know that NASA managers considered design problems in the booster rockets "an acceptable risk," signing waivers to allow each of the last five shuttle launches. Or that *Challenger*'s total weight of 4.53 million pounds was the heaviest in shuttle history, too heavy to reach emergency landing sites in Spain and Germany, and so heavy that NASA needed to sign another waiver to allow it to return to the Kennedy Space Center in case of a launch emergency. Or that John Young, the chief of the astronaut office, had asked several weeks earlier for a safer shuttle, one with "a cockpit abort mechanism" — an escape capsule.

Four days before the *Challenger* launch, Charles Redmond, a NASA spokesman, said the chances of a shuttle accident were no greater than the likelihood of a plane crash.

"But the odds of an accident at Three Mile Island were about a million to one, and the odds of being struck by lightning are about three million to one," he said. "Both of those have happened."

at 1 in 100,000; an independent federal review panel placed the odds at 1 in 1,000. An Air Force study said they could be as low as 1 in 35. Christa knew only that she faced the possibility of having "not a good day."

Neither did she know that NASA managers considered design problems in the booster rockets "an acceptable risk," signing waivers to allow each of the last five shuttle launches. Or that *Challenger*'s total weight of 4.53 million pounds was the heaviest in shuttle history, too heavy to reach emergency landing sites in Spain and Germany, and so heavy that NASA needed to sign another waiver to allow it to return to the Kennedy Space Center in case of a launch emergency. Or that John Young, the chief of the astronaut office, had asked several weeks earlier for a safer shuttle, one with "a cockpit abort mechanism" — an escape capsule.

Four days before the *Challenger* launch, Charles Redmond, a NASA spokesman, said the chances of a shuttle accident were no greater than the likelihood of a plane crash.

"But the odds of an accident at Three Mile Island were about a million to one, and the odds of being struck by lightning are about three million to one," he said. "Both of those have happened."

Christa knew nothing about the odds when she accepted an offer from Corroon & Black Inspace, Inc., of a $1 million life insurance policy to cover her while aboard the shuttle. Robert Berman, a NASA lawyer, had called Christa when she was in quarantine to tell her about the offer. She had accepted it as a goodwill gesture. She had not imagined she would need it.

By Saturday morning, January 25, fifteen thousand special guests, more than twice the usual number, planned to witness the next day's launch. Among them were Senator Jake Garn of Utah, Representatives Bill Nelson and Donald Fuqua, both of Florida, Governor John Sununu of New Hampshire, syndicated columnist and space enthusiast Jack Anderson, NEA president Mary Hatwood Futrell, AFT president Albert Shanker, the Moroccan ambassador to the United States and a delegation from the People's Republic of China. The actor Tom Selleck had declined Judy Resnik's invitation, but Vice President George Bush planned to be the first White House observer of a manned space launch since Richard Nixon watched Apollo 12 lift off in 1969.

The weather was the only obstacle. NASA

issued a bulletin at 10:00 A.M. warning that "an approaching cold front is threatening to deliver rain showers and thunderstorms to the area Sunday morning." Raindrops became bullets when they struck an object traveling at several thousand miles an hour, so NASA had no intention of risking a launch in such conditions. But it planned to wait as long as it could — until a final review meeting that night — before deciding whether to delay the mission. It had a schedule to meet.

Despite the dour forecast, everyone proceeded as if *Challenger* would fly in the morning. Reporters sat through the usual series of day-before briefings on the mission's payloads and purposes. Christa studied the flight plan, underwent a routine medical exam and attended crew briefings on the status of the shuttle, its cargo and the weather. Steve tried to prepare the kids for the tumult of the liftoff by taking them to a movie, *The Dream Is Alive*.

At five o'clock, the crew's bedtime, a couple hundred of Christa's friends and relatives gathered for a bon voyage party. NASA tradition required the spouses of crew members to host a prelaunch celebration, so Steve, June Scobee and Marcia Jarvis hired bartenders, ordered light hors d'oeuvres and rented a ballroom at the Holiday Inn in Orlando. The sky was overcast, the mood festive. The launch had yet to be delayed.

The Corrigans stood by a display of Christa's life in pictures — Christa as a wide-eyed baby, a pig-tailed child, a Girl Scout, a big sister, a bride, a mother, a teachernaut — and greeted people at the door. They hugged old friends and introduced new ones to Christa's four younger brothers and sisters, each of whom had graduated from college.

"They followed Christa's lead," Ed said proudly, "and weren't we lucky to have her leading."

Grace had spent hours helping Steve organize the party. She had worried about whether people would enjoy themselves, but as the countdown continued she was more concerned with Christa's mission.

"I'm getting more and more apprehensive by the minute," Grace said.

Steve stood outside by a small fountain, a *Challenger* crew pin on his lapel, granting brief interviews to several reporters from New Hampshire, his last before the launch. He marveled at the casual mood of the crew and reaffirmed his faith in the shuttle's safety. He said Caroline loved Disney World and Scott loved everything but the press. He said his own experience would be more pleasant if he could relax. He was edgy. Christa had been sleeping better than he, he said, and she had felt none of the tension that tormented him. Still, when a reporter asked him what his last words to her would be before the launch, he said, "Don't quit."

Back in the ballroom, he mingled with Christa's childhood friends, joked with his brother, Wayne, and tried to keep track of Scott and Caroline, who had found friends in the Scobee and Jarvis parties. He toasted the launch with neighbors from Concord, posed for pictures and signed a few autographs, which puzzled Caroline. She thought only her mother signed autographs.

"Daddy, why are you doing that?" she asked, dragging a little white purse behind her.

"Because I married Mommy," he said.

Scott stayed near the hors d'oeuvres and shied from the press — except to predict the Patriots would win the Super Bowl. Later, Caroline conducted a series of television and newspaper interviews in the hall outside the ballroom. None of the questions bothered her; she had heard most of them before.

"What's your mom doing tomorrow?"

"Going up in space."

443

"Do you wish you were going, too?"

"Yes."

"Why?"

"I want to float like her."

"Would you like Scott to go, too?"

"Yes, and Daddy, too. I wish all of us could go."

Rumors on the status of the next morning's launch flared and fizzled through the night. The Corrigans said the launch had been pushed back to the afternoon. Others said it was on for the morning. A few said the approaching cold front had delayed it several days. Finally, at 10:00 P.M. NASA announced it had scrubbed the lift-off until Monday morning at 9:37, the third delay in five days.

The news came as no surprise to Steve, who was encouraged by Jesse Moore's message to reporters afterward. Moore had decided on the delay.

"We're not going to take any kind of risks because of our schedule pressure," he said. "We'll sit on the ground until we all believe it's safe to launch."

Christa's publicist, Linda Long, called her with the news.

"It's a scrub," Long said.

"Make sure you tell my parents so they don't get up at four A.M.," Christa said.

"I'm sorry," Long said, sensing Christa's disappointment. "I know how you feel."

"No, you don't," Christa said. "I'd rather be floating around up there."

The delay prompted hundreds to head home, including most of the class of 51-L. The launch was to be the centerpiece of their reunion, a celebration of renewed respect for education, and they had sent Christa a message inspired by her acceptance speech at the White House seven months earlier. Terri Rosenblatt had written it.

"When that shuttle goes," the message said, "there may be one teacher, but the soul and spirit of millions of teachers and students will be with you."

Soon the vice president announced he could not attend a Monday morning launch. Neither could Garn, Nelson, Fuqua or the Moroccan ambassador. The only blessing in the delay was that two and a half million students and their teachers could now watch the lift-off in their classrooms. It would be a boon for education and for NASA.

It was, after all, the children's mission. Every educational publication from *Weekly Reader* to the Universal Press Syndicate's Sunday *Mini Page* had featured Christa's odyssey. NASA had distributed a half million

445

curriculum guides to teachers across the country, and the 8 teacher-in-space finalists each had traveled as many as twenty-five thousand miles to hype the flight. The 103 semifinalists had crisscrossed their states as space ambassadors, and the nation's teachers' unions had reminded their membership of the mission's value. The teachers had responded, even those who wondered why Christa would accept such a risk.

"I would never volunteer to do what she has done," Karen Rowe, an elementary school teacher in Georgetown, Massachusetts, told her students. "But I love her courage. She's taking a chance for us who chose the safer path."

In Oakland, New Jersey, fifty junior high school students had built a shuttle simulator and planned to go along for the ride, some of them living in the shuttle around the clock, others monitoring the flight by ham radio from "tracking stations" in classrooms about the school. In Framingham, Massachusetts, fourth graders had cut out seven block white letters — C-H-R-I-S-T-A — and taped them to their classroom window with twenty-two handmade stars, one for each of them, symbolizing their intention to reach for the stars. In Boston, students at the Horace Mann

446

School for the Deaf had subscribed to Christa's hometown paper and covered the walls with articles about her adventure. In Manchester, New Hampshire, fourth graders had written personal messages to the teacher-naut, among them Amy Heath's untitled poem:

> *Christa! Christa! Christa!*
> *Have a nice flight.*
> *I'm sure it will be*
> *A wonderful sight.*
> *Christa! Christa! Christa!*
> *Now you're going into space*
> *To be an astronaut.*
> *I can't wait to see your face.*

Nowhere had children prepared more keenly for Christa's mission than in Concord. Robert Veilleux, the state's other teacher-in-space finalist, had delivered a crash course in space science to the city's elementary school teachers. Two NASA education specialists had spent a week visiting each of the schools, and the director of the district's food service had planned to serve the same lunch to students that Christa ate on the day she taught from space — ham and cheese, pineapple chunks, walnuts and a strawberry drink. Only nutritional requirements prevented him from substituting the strawberry drink for milk and walnuts for the healthier GORP (Good Old Raisins and Peanuts).

First graders learned the names of the planets. Second graders used string, straws and balloons to study the fundamentals of rocketry. Nearly three hundred students at the Walker School, where Christa had attended her first interview for the teacher-in-space contest, created a giant relief mural of the shuttle orbiting the Earth through stars that each of them had made. Junior high students used seventy gallons of ice cream to sculpt a four-foot-high space shuttle sundae. Students in Eileen O'Hara's women's history course studied a new history maker — Christa McAuliffe — and children throughout the school district received bookmarks bearing Christa's name, a drawing of a shuttle launch and the message "Reach for the Stars — Read!"

In a special supplement the *Concord Monitor* had published for children across the state, Christa told them the teacher-in-space program was "new and fun, and remember I'm getting ready for the ride of a lifetime!"

Had the weather not played a hoax, the ride might already have begun. Slowed by a low pressure system over southern Georgia, the approaching cold front had arrived late, and at the scheduled launch time on Sunday morning the sky was a rich blue. The wind was calm and the birds were singing. The conditions were nearly perfect.

"All I could think of was: too bad we didn't know this before . . . " said Joe McRoberts, a NASA spokesman.

The *Union Leader* newspaper in New Hampshire summed it up in a front-page headline: NASA: WE GOOFED.

At noon, when Christa might have been "floating around up there," the front finally raced through with rain and cold winds. She ate lunch in quarantine with Steve, then spent the afternoon with the crew autographing stacks of pictures for their families and friends. The sky cleared later and Jarvis asked if she wanted to go for a bike ride. He knew where they could borrow a couple of ten-speeds.

Christa had time to kill before the Super Bowl, so she pulled on a sweater and joined him, pedaling about the flat roads that snaked through the vast space complex, the first astronauts to engage in such an activity on the eve of a launch. A crew from an Orlando television station spotted them and pulled up beside them in a van, their camera whirring.

"Stay six feet away!" Christa shouted, smiling. "I'm in quarantine!"

The television crew followed them awhile, and before they drove away, Christa said, "It must be a slow news day."

Real slow. Only the biggest sports event of the year and the most important space discovery of the decade — *Voyager*'s close encounter with Uranus, the first by mankind. None of the nine hundred journalists at the space center attended a closed-circuit press conference that day to hear scientists describe the seven new moons *Voyager* had discovered around Uranus, but when they learned a television crew had spotted Christa on a bicycle, they fairly trampled each other for details.

Before bed that night, Christa watched the Super Bowl, which helped put her to sleep — the Patriots were flogged, 46–10 — and cost her five dollars in a betting pool she and Steve had entered with a group from Concord. Steve watched the game with Scott, Caroline and friends at Disney World while the Corrigans hosted a Super Bowl party of their own. As the Patriots unraveled, the Corrigans quipped that Christa's flight would help New Englanders erase the pain of the loss.

When a reporter asked Betsy if she was worried about Christa's safety, she said she was more worried about flying to Florida from California.

"NASA doesn't compromise at all," Betsy said. "They make sure everything is correct and one hundred percent before they go."

In case it was not 100 percent, workers collected the crew's wallets and personal belongings the next morning and stored them with visas, ready to send them with a seventy-member rapid response team to a country where the shuttle might ditch in an emergency. A one-hundred-member recovery crew lined the emergency landing strip at the space center. A rescue team stood by in Dakar. It was the eighteenth anniversary of the fire that killed three NASA astronauts on a nearby launch pad.

At 6:45 A.M. the *Challenger* crew strode confidently from the Operations and Checkout Building, a banner pinned behind them intended for Jarvis — HAVE A GREAT FLIGHT, GREGO!

"How do you feel, Christa?" photographers shouted.

"I feel great," she said, smiling, waving, finally wearing flight boots.

Jo Ann Jordan, one of her close friends, watched on television.

"I really thought we'd see more apprehension," she said. "I thought, 'Here comes a lady who's a mother, and she's leaving those babies at home.' "

Anne Malavich was not surprised by Christa's determination.

"Someone else with children the age of Christa's might have said, 'Maybe Caroline's too young; maybe I shouldn't do it,' " Malavich said. "But Christa decided this was something she wanted to do and she went right after it. She knew she had a good support system."

Christa was confused by those who would balk at such an opportunity.

"You know, people come up to me all the time and say, 'I really admire what you're doing, but I wouldn't want to do it,' " she had told a reporter. "I can't understand that. If you had a chance, wouldn't you want to do it?"

Now the wind was brisk and the temperature 40 degrees, chilly for Florida, but a full moon shone in a cloudless sky. It seemed another perfect day for a launch. In the white room, a smiling technician greeted Christa in a black mortarboard and tassel. She laughed, slipped into her safety harness and crawled through the circular hatch to the shuttle at 7:36 A.M. Then she waited.

Steve and the children watched with the immediate families of the other crew members at the Launch Control Center, four miles away. The Corrigans, who were not considered immediate family, waited on aluminum bleachers in a nearby VIP viewing area, the top half of *Challenger* visible on the horizon. Ed wore a stadium coat and a Scottish driver's cap, Grace a white coat with a fluffy collar. They looked through a set of giant binoculars and waited, their four other children beside them.

Thousands of people had left, but thousands remained, and the scene at the space center was similar to the sidelines at a Thanksgiving football game in New England. Huddled under blankets in twos and threes, hoods on their heads, scarves around their necks, hot chocolate and coffee in their hands, most of them watched the clock and waited. Others ran in place, shivering. A few, prisoners of their Yankee pride, raised their chins to the wind.

"Just like home," said a coatless Michael Metcalf, the teacher-in-space finalist from Vermont, his nose red from the cold.

Scott's classmates, many wrapped in blankets, waited with their chaperones and a small army of journalists on the edge of a parkway across a swamp from the launch pad. They read, played games and threw rocks into the swamp, hoping to spot an alligator. They waited.

An hour before launch time, the children learned their first lesson in technology's imperfections. Ron McNair spotted a microswitch on the circular hatch that indicated the hatch was not closed. He radioed Scobee, who asked technicians to check it. Using test equipment in the white room, they determined the hatch was closed and that the microswitch must have malfunctioned. But McNair was not convinced and he relayed his doubts to Scobee, who insisted the white-coated technicians open the hatch and check it again, a routine procedure that turned ridiculous.

A quick inspection confirmed the microswitch had failed, but when the technicians tried to close the hatch again, the door handle refused to budge, frozen by a stubborn four-inch bolt. They wrestled with the bolt awhile, then called for a drill and hacksaw. Forty-five minutes later, the tools had yet to arrive, and the angry voices of the restless technicians were audible on NASA's loudspeakers and television network. When the delivery man arrived minutes later, he gave them only a drill, whose battery died before the bolt would budge. They ordered a more powerful drill and, again, a hacksaw.

Meanwhile, the winds kicked harder.

"I'm mad," said Sarah Carley, one of Scott's classmates. "We came down here to see that shuttle take off and if it's not going to take off, I'm going to have a hyper-spaz." A hyper-spaz, she explained, is "when someone goes crazy and takes a fit."

But Scott's best friend, Zachary Fried, waited patiently. Space was his consuming passion, and with the other children he had heard astronaut Vance Brand explain two days earlier that the shuttle was a marvelous machine "when everything works, but only when everything works." Zachary was a serious boy, and he sensed the dangers. He was willing to wait.

The mood was mixed at Concord High, where journalists seemed to equal the twelve hundred students in number. Seated before televisions throughout the building, some of the students wore party hats decorated with pictures of the space shuttle. Some wore T-shirts that said "Concord — Where the Spirit Is High." Others tooted tiny noisemakers or played the military bugle charge on kazoos. A few raised banners that said WE'RE WITH U

455

CHRISTA and TO BOLDLY GO WHERE NO TEACHER HAS GONE BEFORE, paraphrasing the *Star Trek* television theme. The yearbook staff had recently designed a cover for the 1986 edition that featured the space shuttle orbiting the school seal. The theme was "Christa in Space."

But two students walked about the building in protest, toting handmade signs that said I'D RATHER BE LEARNING, complete with "shuttle-buster" logos.

"I'm really getting sick of this," said one of the protesters, Andrew Cagle. "All we hear is Christa this, Christa that."

Andrea Rice, a classmate of Cagle, was more sympathetic. "I pity Mrs. McAuliffe," she said. "She must be, like, losing it."

Actually, Mrs. McAuliffe was sleeping. She had joked for a while with Jarvis about the hatch problems, but after several hours of lying flat on their backs, strapped onto seats that felt like thinly padded, cold steel kitchen chairs, staring at nothing but the middeck lockers, they began to lose the circulation in their legs and their humor as well. Jarvis's legs went to sleep and Christa dozed off altogether.

She awoke when the technicians attacked the titanium bolt with the second drill, which proved as useless as the first. The bolt was so hard that it chewed up the drill bit. Christa listened as they finally removed the bolt with the hacksaw. It was 11:07.

"Isn't this ridiculous?" Grace Corrigan said.

"I would have gotten the hacksaw sooner," said Ed.

"I would have gotten my nail file," said Grace.

The winds had begun gusting to thirty miles an hour — too violent for the shuttle to attempt a landing at the space center in case of an emergency — and the flight soon was postponed another twenty-four hours.

"I guess we'll have to wait another day, but it's getting emotionally grating," Grace said, turning her back on *Challenger*.

Her son Steve and daughter Betsy could stay no longer, and she worried, too, for Christa. The delay was the fourth in six days, the sixth overall, one short of tying the record. The cost to the taxpayers: $200,000.

ANOTHER SHUTTLE CHANCE BOLTS AWAY said the headline in the *Washington Post*.

"It was just not our day," said Robert Seick, the director of shuttle operations.

Indeed, it was so bad that Steve tweaked the technicians by sending them a tool set. And so bad that Christa lost her spirit. After five and a half hours of uneasy confinement, she slouched back to the crew quarters past the same photographers she had waved to in the morning. This time there was no wave, no smile, no bounce to her step. There was a look of deep frustration.

"Her face really struck me," said Mark Beauvais, the school superintendent who knew her well, whose wife had taught with her, whose daughter had babysat for her. "It was a look I'd never seen before. It bothered me."

Christa ate an early dinner with Steve that night at the crew quarters, kissed him good-bye for the last time and called Eileen O'Hara in Concord. She wanted her students to know how she had felt during the long, yawning wait.

"Go borrow a motorcycle helmet," Christa said. "Lie on the floor with your legs up on the bed, and lie there for five hours. You can't read, you can't watch television, you can't have anything loose around. You're strapped down really tightly, with oxygen lines and wires coming out of your suit. You can hardly say anything. Just lie there and you'll know how it feels."

But Christa managed to laugh about the bolt. Here was NASA with a billion-dollar shuttle, the most complex flying machine in the world, and not a tool box to be found. She suggested it might make a bad television comedy. Then she talked about her family. Christa was pleased for her mother that the reception had been a success. She was proud Caroline had been a leader at the party by engaging several other children in games. She was sad Scott had endured intense media pressure, and she was excited about the chance that she would fly in the morning.

"I still can't wait," she said.

Neither could NASA. A record cold front that threatened to destroy central Florida's citrus crop was due to rush in with subfreezing temperatures and Arctic wind-chill factors that would last through the next morning's 9:38 launch time. No shuttle had been launched at a temperature below 51 degrees, but the space agency was about to test *Challenger*'s limits, to take chances it had never before taken. Pressure to meet its schedule was intense, some of it self-imposed, some of it, NASA officials argued, created by the media.

"Yet another costly, red-faces-all-around space shuttle launch delay," Dan Rather reported that evening. "This time a bad bolt on a hatch and a bad-weather bolt from the blue are being blamed. What's more, a rescheduled launch for tomorrow doesn't look good, either. Bruce Hall has the latest on today's high-tech low comedy."

The press deserved 98 percent of the blame for the pressure, argued Richard Smith, the director of the Kennedy Space Center. Journalists did not take part in the decision to proceed with the launch, however. Smith did. For the first time in the history of the shuttle program, NASA decided to load the external fuel tank in subfreezing weather.

"We might learn a lot of interesting things," said George Diller, an agency spokesman.

Engineers for Morton Thiokol, the company that manufactured the solid rocket boosters, cautioned that cold temperatures could cause the O rings — the rubber seals between the rocket's sausagelike segments — to lose their resilience and allow explosive gases to leak, an event that could be catastrophic. When NASA officials at the Marshall Space Flight Center downplayed the warning, one of the engineers, Allan McDonald, replied, "If anything happens to this launch, I sure wouldn't want to be the person to stand in front of a board of inquiry to explain."

The engineers were overruled by their superiors, and the giant digital clock continued to tick. Antifreeze was pumped into fluid lines on the launch tower, and the tower itself had grown a thin coat of ice by 10:00 P.M.

"The launch crew is on schedule," reported NASA spokesman James Mizell. "They say it's a miracle because the winds are gusting very hard."

As Monday turned to Tuesday, January 28, the ground crew evacuated the launch pad and began the three-hour process of loading the external tank with a half million gallons of super-cold, highly explosive propellants. Twenty minutes later, the fueling stopped. A computer had detected an electrical problem in the fire detection system, a problem that took two and a half hours to solve. The launch was pushed back to 10:38 A.M.

"We're down several hours," said Hugh Harris, the public affairs chief, "but that doesn't mean we can't catch up some of that time."

As the *Challenger* crew slept, the team at the launch pad grew increasingly wary of the ice. The temperature had dipped to 24 degrees, the wind chill to 10 below zero and, like northern homeowners, the ground crew had kept the water running to prevent the pipes that fed the fire extinguishing system from freezing. Still, a pipe froze, and the cold caused the failure of eleven cameras that

461

NASA had placed in the area to record the lift-off. The launch pad was colder than parts of New Hampshire.

"Some damn Yankee came down here and forgot to shut the door," shouted a shivering reporter into the press center before dawn.

When Christa and the crew woke at 6:20, the full moon spattered silver on the choppy waters of the Atlantic. From her window Christa saw the orbiter glitter in the distance, forty-five flood lamps filtering through the steel launch tower. She showered, pulled on blue jeans, a white crew shirt, her powder-blue sneakers and joined the crew for a light breakfast, her hair still wet. Christa sat at a rectangular table, a centerpiece of roses and American flags before her, Ellison Onizuka to her right, Mike Smith, in his stocking feet, to her left. Bleary-eyed, they smiled when the cooks placed a white-frosted cake on the table. The cake was decorated with their crew emblem, the shuttle *Challenger* and each of their names.

At 7:20, they received their final weather briefing. They were told about the 25-degree temperature, the wind chill in the single digits and the ice at the pad. The day had dawned a pearly white, and it seemed better suited for riding a sleigh than a shuttle, but the winds had begun to weaken and the sky was clear. They were encouraged. None of them knew the dangers of a cold day.

"My kind of weather," Dick Scobee said. "What a great day for flying!"

A few minutes later, Scobee led the crew down the cement ramp past the roped-off throng of photographers, their steamy breath rising in unison. Resnik walked at his side. McNair appeared next, then Smith, who rubbed his hands together.

"Wo-o-o-o-o!" he said, shivering.

Like Resnik, Christa wore yellow gloves. She had rolled up her sleeves because they were too long. Her eyes were bright and the bounce was back in her stride. Christa was going to work. She was going up in space. She was smiling.

"Christa, hey, Christa!" the photographers shouted.

"We're going to go off today," she said hopefully, waving good-bye.

The cold pinched her cheeks as she crossed the catwalk. She saw icicles several feet long on the launch tower, and the giant external tank seemed to groan as its aluminum skin contracted from the chill of the fuel and the air. An unusual number of ground-crew workers huddled in the white room for shelter.

"We *are* going to go off today, aren't we?" Christa wondered aloud.

Scobee jumped around to keep warm, then donned his skullcap, helmet, safety harness and crawled into the orbiter. Smith followed, then Resnik, who hugged one of the technicians before boarding. As McNair inspected the hatch that had grounded them the day before, Johnny Corlew, a quality assurance inspector, gave Christa a gift — a red Rome apple.

Corlew, who had grown up in Indiana, had picked apples for his teachers from the tree in his family's yard. When he learned he had been assigned to the white room for the teacher-in-space mission, he had asked his wife to buy him one at the supermarket for Christa.

"Save it for me," Christa said, smiling, "and I'll eat it when I get back."

At 8:35, she shook Corlew's hand and entered the shuttle. Jarvis and McNair followed, then Corlew closed the hatch, drawing a round of applause from the rest of the ground crew. At last something had gone right.

Launch Control radioed the crew seconds later to test their headsets.

"Let's hope we go today," the controller said to Scobee.

"We'd like to do that," he said.

The temperature in the cabin was 61 degrees, cooler than Christa's apartment at Peachtree Lane.

"Brrrr," the controller said.

"Brrrr, is right," said Scobee.

At Disney World, Steve had scraped the frost from the windshield of his rented car, gathered Scott and Caroline and headed to the space center, doubtful they would see a launch, indeed hoping they would see one only if the conditions were perfect. The conditions did not seem perfect.

"Good morning, Christa," the controller said, testing her headset. "Hope we go today."

"Good morning," she said in a teacher's singsong. "Hope so too."

Except for the "loud and clear" of a later radio test, they were her last known words.

At 9:08, a half hour before the original launch time, the countdown stopped unexpectedly. Rocco Petrone, the president of Rockwell Space Transportation Systems, a division of Rockwell International, which built and managed NASA's shuttle fleet, had seen television pictures of the icicles on the launch pad. Concerned that the icicles might snap during the lift-off and damage the tiles that protect the orbiter during reentry, he called the space agency to advise against a launch. NASA dispatched its ice inspection team.

Announcing the delay, launch director Gene Thomas told the crew they sat on "the northernmost pad we've got, and you can probably tell from the icicles out there how far north it is."

Then quickly, reassuringly, he added, "We're going to give you a ride today."

He was right. An hour later NASA managers determined the conditions were safe and ordered the countdown to resume, this time to a launch at 11:38. In the press center, a radio reporter from New Hampshire shouted into his microphone, "We are approaching the moment of truth!" Several rows in front of him, William Broad, the space writer for the *New York Times*, a hood tied snugly under his chin, hunched over his computer keyboard, typing with gloved hands. He thought it a strange day for a launch.

The news sparked a brief celebration in the viewing area, where spectators had waited as if the *Challenger* were an expectant mother. Their cheers might not have died so quickly in the breeze had more people stayed. Monday's delay had prompted thousands more to head home — Steve and Betsy Corrigan, Mike Smith's brother, Greg Jarvis's mother,

Christa's college friends, her teaching colleagues from Maryland, her Girl Scout companions, many of her friends from Concord, the governor of New Hampshire and his son, the delegation from China, all but about twenty of the class of 51-L, hundreds who had come to see the teacher's launch, hundreds who had come simply to see a launch. Even at the space center campground, where dozens usually jammed the roofs of Winnebagos to whistle and wave to departing astronauts, only a faithful few remained.

The VIP bleachers were so nearly empty that NASA had bused in Scott McAuliffe's classmates from the parkway. Wearing red, white and blue baseball caps and bundled in every layer of clothing that fit, they stood with Christa's parents, who had left a more secluded viewing area to join the children as a symbol, they said, of Christa's commitment to education. Her sister Lisa and brother Christopher stood there as well. Christopher stirred a cup of cocoa for what seemed like minutes, his eyes fixed on the idle *Challenger*. Nearby, Steve's sister, Melissa, held her three-year-old daughter, Shana.

"Christa's just perfect for this," she said, excitement in her voice, her cheeks pink from the morning chill.

"Mommy," Shana said, "I want to go too."

Barbara Morgan sat a hundred yards away on the windblown roof of a shack next to the press grandstand, preparing to beam her "Mission Watch" program into the nation's classrooms. She dug her hands into the pockets of a light windbreaker, shivering while she waited for someone to bring her a sweater. She had delivered a bon voyage card to the crew two days earlier, and when Monday's launch had been scrubbed, she had reminded everyone, "We have to think about safety, but it's still going to be a wonderful mission."

Now it was 11:29. The final planned hold had ended and the giant digital clock had begun to tick again. The mercury had risen above the freezing point.

On a highway in Daytona, Joanne Brown, who had worn Christa's wedding dress in her own wedding, pulled over to watch the launch. Near Jacksonville, Patricia Mangum, whom Christa had taught to play the guitar, watched from a roadside with Christa's colleagues from her first years of teaching. One of their young daughters looked skyward through the red binoculars Santa had given her. In East Texas, Les Johnson, the chef from Pe-Te's who had introduced Christa to Cajun food, listened to the countdown on his car radio. At Steve's red-brick law offices in Concord, fifteen of his co-workers huddled about a television.

Eileen O'Hara sat surrounded by reporters before a television in the Concord High auditorium where Christa had said good-bye six months earlier. Students blew noisemakers and sounded the charge. Among them were Andy Bart, who intended to ask Christa by satellite how the world looked without boundaries, and Rick St. Hilaire, a space buff who planned to ask her how the toilet worked. O'Hara wore a shuttle pin Christa had given her and assured everyone that "so many thousands of people are checking everything" to ensure a safe launch. A balloon bobbed above her.

At a kindergarten five miles away, Rosemary Martin told her students they were not about to watch a scene from *Star Wars*.

"This is not a fantasy," she said slowly. "It's real, and it's extra special because Christa McAuliffe is just like Mrs. Martin. And Mrs. Martin wishes she could go, too."

At the Walker School, Josh McLeod and Nissa Barker, both six years old, moved closer to the television. They each wore two cardboard boxes covered with aluminum foil —

one a space suit, the other a helmet that had holes for their eyes. Nissa wore an airpack she had made from a pink shoebox and had decorated with a yellow sun, a blue Earth and a red moon. She said she wanted to ride the *Challenger* as well.

"I would be a little bit scared," she said, "but mostly excited."

Not since the glories of the moon landings had space lured so many. From the Virgin Islands to an Eskimo village in the Arctic Circle, they waited — two and a half million students and their teachers, among them Christa's colleagues and most of the class of 51-L. They watched at the Sally Ride Elementary School near Houston, the McCall-Donnelly Elementary School in Idaho, the Thomas Johnson Middle School in Maryland. In Framingham, Charlie Sposato's English students continued working through the countdown. They studied *2001: A Space Odyssey*.

In the White House, Nancy Reagan sat before a television, awaiting the launch. The president was busy at work. He was scheduled to deliver his State of the Union address that night, a speech in which he would trumpet the success of the space program as the first private citizen circled the Earth.

Aboard *Challenger*, Christa lay on her back for the third straight hour. She had nothing to read, to see or to do but plenty to consider — her family, her students, the strange road that had taken her there, the unknown ride ahead of her, her mystery vacation with Steve afterward. One thing she did not consider were the first words she would relay back to Earth.

"I don't want to say anything phony or contrived," she had said earlier. "I want my perceptions to be honest. I want them to be just as natural as possible, and if they're very ordinary, well, that's okay. Maybe that's just me."

The visor on her airtight helmet had snapped shut. She was breathing pure oxygen. Beside her sat Greg Jarvis, who looked upon the flight as "the final act of a well-rehearsed play."

As the countdown dipped below three minutes, Mel Myler, the director of the New Hampshire chapter of the National Education Association, moved closer to his young son, who had recently sent Christa a note. The note said, "Thanks for getting me involved in the space program. Love, Jason."

Now they were chanting with dozens of others, "Gimme a *C* . . . gimme an *H* . . . gimme an *R* . . . "

At the Apollo Elementary School, the Sea

Missile Motel, the Challenger Lounge, the Moon Hut Restaurant on Astronaut Highway, all along the space coast people stopped what they were doing and turned to the sky. Braced for the cold, reporters by the dozen filed out of the press center to their tables at the outdoor grandstand. Many of them had applied to ride with Mike Smith in September as the first journalist in space. Brian Ballard, the editor of the Concord High newspaper, stood with them, a video recorder in his right hand, a 35mm camera in his left, ready for the triumph that would help him forget the shooting tragedy that still haunted his school. Nearby, Linda Long, who described Christa as a twenty-first-century Cinderella, pulled a video recorder from her camera bag.

"Please, God," she said softly, "take care of Christa and the crew."

"Ninety seconds and counting," said the voice on the loudspeakers. "The 51-L mission ready to go."

Scott's classmates clapped in rhythm. The chaperones focused their cameras and binoculars. A banner behind them said GO CHRISTA!

"Have a good mission," launch director Gene Thomas told the crew.

"Thanks a bunch," Scobee said. "We'll see you when we get back."

At T-minus twenty-five seconds, *Challenger*'s computers ran through their final checks. Sea gulls swirled about the pad. Waves rolled against rocks on the shore. The crew waited. The ride Thomas promised them was about to begin.

The children counted in unison: "Ten, nine, eight . . . "

Christa's sister Lisa stepped back a row to stand with her parents. Grace clutched Lisa's shoulder with her right hand and pressed the fingers of her left into Ed's back. Ed pulled her close, and as he reached for Lisa's hand, he imagined Christa holding hands with Greg Jarvis, her eyes open wide. Christopher removed his teacher-in-space cap for the last time.

With six seconds to go, the main engines started and the shuttle disappeared in a steam cloud that rolled a mile in every direction. A NASA engineer held his breath.

"Bye, Christa," Barbara Morgan said, smiling and waving. "Bye, Christa."

On the roof of the Launch Control Center, Scott and Caroline stood at Steve's side, watching. Steve looked through the lens of his video camera, waiting. Then, at 11:38 A.M., the solid rocket boosters ignited and *Challenger* shook the Earth good-bye.

EPILOGUE

Six months after *Challenger* lifted off, Christa's remains lay in an unmarked grave overlooking a winding river and the evergreen hills near her New Hampshire home. Two freshly planted maples framed the burial spot, a pot of red geraniums blooming between them. A few tiny American flags snapped in the summer wind, and hidden in the grass beneath the flags was a lapel pin depicting the launch of a space shuttle. It said, "I want to GO."

Those were the only visible signs that Christa had returned to Concord from shuttle mission 51-L. Steve wanted it that way. He had watched *Challenger* burst from its giant steam cloud and rocket skyward at twice the speed of sound. He had seen it explode seventy-three seconds into flight, sending Christa and her six companions to their deaths. Then he had fled into seclusion, sad, lonely and angry, trying to cope with the tragedy and to protect his private memories of a wife whose life and death had become so public.

Steve had leaned on Christa. She had inspired him since the September morning he fell in love with her in Sister Seretina's homeroom. She had nurtured and counseled and shepherded him for so long that while her absence in the months before the *Challenger* mission helped him prepare for life as a single parent, nothing had prepared him for living without her.

Steve stayed with Scott and Caroline in Florida in the days after the explosion, trying to help them understand what had happened before they returned to a home filled with reminders of their mother. Scott soon accepted Christa's death, but Caroline needed time. She remembered Christa coming and going so often in recent months that she believed Christa would return again. Caroline was confused, and not until she went to Disney World with the wife of one of Steve's law partners a couple of days later did she fully understand her mother wasn't coming back. She was sad. She asked if she could ride on Dumbo.

The next morning, on the day Christa was to have taught from space, Steve flew from Florida to the Johnson Space Center for one of his few public appearances after *Challenger* went down. He sat before a building where

Christa had trained, his head bowed as he listened to the president eulogize the *Challenger* crew. The Corrigans, Barbara Morgan and the class of 51-L sat in the crowd behind him. The families of Christa's crew mates sat beside him, among them eight-year-old Erin Smith, clutching a teddy bear with a pink apron, and eighteen-month-old Joy McNair, a pacifier in her mouth and a red bow in her short black hair. The baby looked skyward when the 539th Air Force Band played "God Bless America," Staff Sgt. Susan Arnold's silver trumpet falling silent as four T-38s burst in formation out of the northern sky, one of them disappearing into the clouds — a symbol of the lost crew.

"The air was so full of sorrow," Arnold said. "I had no music left inside of me."

The music started in Concord soon after it stopped in Houston. The twelve hundred students at Concord High, dozens of whom had sought counseling to cope with Christa's death, gathered in the school gymnasium to share their feelings in three hours of personal remembrances, poetry and song. The service ended to the soulful melody of "Life in a Northern Town," a ballad about the chilling aftermath of President Kennedy's assassination and the final images of a friend who died too young. Matt Mead, whose college recommendation arrived the day Christa died, sang along. So did a boy who snuck into the service after he had dropped out of school. When the song ended, the boy walked into the principal's office and asked to enroll again. He believed Christa would have liked that.

The city's younger children confronted their grief with crayons. Third grader Brandi Wilson drew four bent tulips beneath a sun that said "I don't feel like shining." Second grader Erin Neville sketched herself crying beneath an exploding shuttle, and her classmate Amy Collard drew Christa waving goodbye in her space suit.

"Know what, Christa?" Amy wrote on her drawing. "You will always live in my heart."

Thousands huddled that night in the bitter cold at the state house plaza to bid Christa good-bye. Many of them had seen Christa there on her last night in Concord. Hundreds had honored her there on Christa McAuliffe Day, and dozens had cheered her there on her triumphant parade ride down Main Street. Christa's victory had been their victory, and now she was gone. They sang for her, sent aloft seven black balloons to honor the *Challenger* crew and then held hands in silence as a

Concord High librarian rang the bells in a nearby church once a minute for seven minutes. As the bells tolled, a police officer knelt in the snow and prayed.

Steve and the children returned to Concord amid a worldwide display of sympathy. The Olympic flame burned again in Los Angeles, candles flickered in chapels across the country and porch lights intended to honor the first teacher in space instead shone as a memorial to her. From Hawaii to New Hampshire, schools, libraries, airports, streets, bridges, mountains, summer camps and holidays were dedicated to Christa and her companions. Millions of dollars poured into scholarship funds. A woman offered to replace Scott's Fleegle with her own stuffed frog named Fleegle. Others asked to make Scott another Fleegle. Within days, tens of thousands of songs and poems and flowers and letters arrived in Concord from as far as China and Poland, among them a letter from Jason Eads, an eleven-year-old in Long Beach, California.

"I'm deeply touched by the death of Christa," Jason wrote to the McAuliffes. "It is no fun. I know how it is. My dad was murdered when I was seven."

In a prison workshop two miles from

Christa's home, a former Concord High student serving a life sentence for murder helped build a memorial to Christa, a wooden plaque inscribed with the message "Reach for the Stars." The plaque stood at the high school in a storage room filled with memorials from as near as New Hampshire and as far away as Japan. It stood for the impact of Christa's death.

"Do you think there will ever be a day I don't think about this?" asked Zachary Fried.

Christa's survivors — her family, friends, colleagues, students and neighbors — faced unsettling news in the weeks after her death. For many of them, sorrow turned to anger amid indications that *Challenger*'s mission failed not because of technological imperfections but human error. Then came more disturbing news. After the survivors believed for thirty-eight days that Christa and her crew mates had died instantly, consumed by a giant fireball, *Challenger*'s cabin and the crew's remains were raised from the floor of the Atlantic, producing enough evidence for NASA to determine that the astronauts had survived at least several seconds after the explosion, possibly until they struck the ocean surface. The anger grew.

A presidential commission blamed the shuttle tragedy on a mismanaged space agency, and Michael Smith's wife, June, filed a $15 million lawsuit (the families of other crew members considered similar action). *Challenger* exploded, the commission reported, because NASA managers ignored months of warnings about the shuttle design and allowed *Challenger* to lift off in weather so cold it caused the O rings between segments of its right booster rocket to fail. Fiery gases burned through the rocket, leading to the catastrophic combustion of hundreds of thousands of gallons of liquid oxygen and liquid hydrogen nearly nine miles high. The only hint the crew sensed trouble was Mike Smith's words a millisecond before the explosion.

He said, "Uh-oh."

Shielded by a willing city, Christa's family buried her in the Catholic section of Concord's public cemetery on May 1, 1986, a year to the day after she expressed her philosophy of living — "to get the most out of life as possible" — for NASA's teacher-in-space contest. Christa's cousin, the Reverend Leary, who performed her wedding mass, conducted a simple graveside service on a hill speckled with the headstones of hundreds of ordinary people. The last snow had melted and the trees had emerged from winter's rest. There were signs of hope.

Strengthened by Grace Corrigan, whose own parents died young, Steve and the children quietly returned to their normal routines. Grace lived with them, cooking, cleaning, gently urging them to press on and get the most out of life. She brought Caroline to a memorial tree-planting in a park where Christa often jogged. She walked Caroline to school, holding her hand and once bringing a bright yellow forsythia for Caroline's teacher. She helped Scott and Caroline paint Easter eggs and hang them from the tree in their front yard. She camped out with her daughter Lisa and Christa's lifelong friends from Girl Scouts. And she spoke in Christa's place at the Framingham State College commencement.

"Christa McAuliffe is a hero, a real hero," Grace told the class of '86, "but perhaps not for the reasons you might think. She is not a hero because of her selection as the first ordinary citizen to venture into space. She is not a hero because she provided us with a wonderful role model. Rather, she is a real hero because long before the teacher-in-space program was ever thought of she overcame many of life's ordinary obstacles and became a worthy person."

Steve returned to work, placed a large photograph of Christa on the wall behind his desk and spent much of his time following the investigation of her death. His colleagues raised a wall of silence around him, and for a while he appeared publicly only to represent his clients or to accept several memorials to Christa. But he slowly eased his isolation.

Steve sometimes rode a bicycle on the quiet streets near his house, Caroline pedaling beside him on her tricycle. Occasionally, he tossed a football with Scott in the driveway, and one day he took Scott to the city's ice rink to skate with Bobby Orr, perhaps the greatest hockey player ever. Orr had traveled to Concord at the suggestion of Scott's godfather, who hoped the visit would soothe Scott's sadness. Orr skated with Scott and his teammates, coaching them, encouraging them, laughing with them. He sat in a quiet room with Scott and Caroline, urging them to visit his house near Boston to see his new puppy. Then he shook hands with Steve and Scott and kissed Caroline on the cheek.

The healing continued, and several weeks later Steve spoke publicly about Christa for the first time since her death. He flew to Louisville with Scott, Caroline and the Corrigans to address the National Education Association's annual convention and accept the Friend of Education Award — the NEA's highest honor — on Christa's behalf. He described Christa to seventy-five hundred educators as "the most selfless person I have ever met," and more.

People considered Christa "a Pollyanna in teacher's clothing, the girl next door and all that stuff," Steve said. "And actually she was. But she was the best kind of Pollyanna. She was a Pollyanna with a strong sense of realism and pragmatism, and not the least important, of politics. Because as you know . . . dreams and ideals are wonderful, but if you can't carry them into action, you might as well not have them."

He urged the educators to continue Christa's dream. He asked them to work to improve the nation's schools.

"If you sit on the sideline, reflect back on Christa as a hero, or as a glorious representative or a canonized saint, rather than putting your energies into accomplishing for her what she wanted to do," he said, "then I think her efforts will have been in vain. You will have turned the teacher-in-space program into a feeble substitute for desperately needed help."

Steve returned to Concord the same day with Scott and Caroline. It was the Fourth of July, two months after Christa had come home to Concord, and that afternoon two tiny new American flags, the kind that children plant, appeared on her grave.

ACKNOWLEDGMENTS

I had the privilege of writing this book because Peter Osnos of Random House had the vision of a story that should be told and trusted an ordinary reporter from a small New Hampshire newspaper to tell it. His guidance in the months after the *Challenger* tragedy was invaluable, as was the assistance of Chief Copy Editor Victoria Klose.

Like the *Concord Monitor*'s coverage of Christa's story, this book is the result of a collaborative effort with my colleagues — executives, editors, reporters, photographers, librarians and telephone operators. I owe special thanks to George Wilson, Tom Haley, Mike Pride, John Fensterwald, Mark Travis, Scot French, David Olinger, Ralph Jimenez, Pam Byrne and Richard Mertens. The contributions of the entire staff were immeasurable.

My thanks also to reporters across the country who helped me see things I was unable to see, especially Martha Cusick of WNHT-TV in Concord, David Tirrell-Wysocki of the Associated Press, Michael Kranish of the *Boston Globe*, Shawne Wickham of the *Union Leader*, Joe Sciacca of the *Boston Herald* and John Getter of KHOU-TV in Houston.

Two NASA public affairs officers — Ed Campion and Barbara Schwartz — deserve special credit for making my job easier from the moment I picked up Christa's trail. I am indebted to many others in and out of NASA who shared their time and thoughts in a particularly difficult time, but it would be redundant to identify them here because their names appear in the book. I could not have written it without them.

Finally, and most of all, I am grateful for the enduring support of my family.

THORNDIKE PRESS HOPES you have enjoyed this Large Print book. All our Large Print titles are designed for the easiest reading, and all our books are made to last. Other Thorndike Press Large Print books are available at your library, through selected bookstores, or directly from the publisher. For more information about current and upcoming titles, please call us, toll free, at 1-800-223-6121, or mail your name and address to:

THORNDIKE PRESS
P. O. BOX 159
THORNDIKE, MAINE 04986

There is no obligation, of course.